D1544744

High-Speed Wireless ATM and LANs

For a listing of recent titles in the *Artech House Mobile Communications Library,*
turn to the back of this book.

High-Speed Wireless ATM and LANs

Benny Bing

Artech House
Boston • London

Library of Congress Cataloging-in-Publication Data
Bing, Benny
    High-speed wireless ATM and LANs / Benny Bing.
        p. cm. — (Artech House mobile communications library)
    Includes bibliographical references and index.
    ISBN 1-58053-092-3 (alk. paper)
        1. Local area networks (Computer networks) 2. Asynchronous transfer mode.
    3. Wireless communication systems. I. Title. II. Series.
TK5105.7 .B56 2000   99-052311
004.6'8—dc21   CIP

British Library Cataloguing in Publication Data
Bing, Benny
    High-speed wireless ATM and LANs. — (Artech House mobile
    communications library)
    1. Asynchronous transfer mode 2. Local area networks
    (Computer networks) 3. Wireless communication systems
    I. Title
    004.6'8

    ISBN 1-58053-092-3

Cover design by Igor Valdman

© 2000 ARTECH HOUSE, INC.
685 Canton Street
Norwood, MA 02062

International Standard Book Number: 1-58053-092-3
Library of Congress Catalog Card Number: 99-052311

10 9 8 7 6 5 4 3 2

To My Mother

CONTENTS

Preface

Wireless local area networks (LANs) provide cable-free access to data rates of 1 Mbps or higher for both indoor and outdoor environments. A substantial portion of the cost of LAN deployment is in interconnecting end-user devices, which many networking experts acknowledge can sometimes exceed the cost of computer hardware and software. A wireless LAN removes the labor and material costs inherent in wiring. It also offers the flexibility to reconfigure or to add more nodes to the network without much planning effort and cost of recabling, thereby making future upgrades inexpensive and easy. The ability to add new mobile computing devices quickly is another main consideration for choosing a wireless LAN. Thus, the proliferation of cheaper, smaller, and more powerful portable notebook computers has fueled tremendous growth in the wireless LAN industry in recent years.

While wireless LANs predominantly support data traffic, a growing emphasis on multimedia traffic has underlined the importance of asynchronous transfer mode (ATM) networks which are capable of supporting demanding applications such as audio/video playback, image browsing, real-time voice/video transmission, and interactive data exchange in a unified manner. However, the extension of ATM from wireline to wireless creates a new set of challenging technical issues. ATM standards are developed based on high data rates and reliable transmission links. This is in contrast to wireless links where bandwidth is limited and error rates are high. Furthermore, existing ATM specifications are designed primarily to provide services to fixed end-terminals with little ability to adapt to mobile connections and the highly time-varying conditions associated with wireless networks.

This book provides a concise discussion on current wireless LAN technology and offers a glimpse of where the field of broadband wireless ATM is heading. Much of the material is based on research conducted by the author. The approach taken in this book emphasizes core concepts and underlying principles rather than factual descriptions. In addition, many carefully prepared illustrations are used throughout to enhance the textual explanations. By distilling details down to the basic issues needed for intuitive understanding, both advanced and novice readers are able to gain valuable insights into the exciting field of high-speed wireless communications and mobile computing. However, the purpose of the book goes beyond making the reader merely proficient in these issues. The many problems discussed serve to stimulate reflection and further research. To encourage readers to fully explore the topics covered, useful Internet resources and references have been included. Most of the references were chosen because they are either informative or particularly well written. Hence, the book can serve as an ideal course reader on advanced wireless networks.

The chapters of this book are organized as follows. Chapter 1 highlights the major technical problems associated with the indoor wireless environment and recommends some solutions to cope with these problems. Various types of wireless LANs are described in Chapter 2. Chapter 3 deals with issues related to wireless LAN implementation while in Chapter 4, the structure of the IEEE 802.11, HIPERLAN, and other emerging industry standards such as Bluetooth and HomeRF are explained. Chapter 5 focuses on the performance evaluation of several commercial wireless LANs, including those complying to the IEEE 802.11 standard. Chapter 6 surveys the technical and service issues related to the deployment of ATM in third-generation wireless networks. The requirements for multimedia communications using ATM are also examined. Several wireless ATM prototypes and products are then described. The chapter concludes with an interesting account of how ATM cells can be transported over satellite networks. Chapter 7 provides a brief overview of the current wireless ATM standardization activities in the United States, Europe, and Japan. Finally, a list of references is included in a bibliography.

Acknowledgments

I wish to express my gratitude to many international colleagues whose technical advice and comments have contributed significantly in shaping the contents of this book. I am very privileged to be able to consult Victor Hayes of Lucent Technologies (Victor chairs the IEEE P802.11 Standards Working Group for Wireless LANs) and Larry Taylor of TTP Communications (Larry is one of the original developers of the HIPERLAN Type 1 specification), both of whom have taken time from their busy schedules to answer some of my queries on the IEEE 802.11 and HIPERLAN standards. Special thanks also go to Jan Boer (Lucent Technologies), Dean Kawaguchi (Symbol Technologies), Larry Ross (Aironet Wireless Communications), and Andrew Crispino (Proxim) for sharing their expertise on the different physical layers of the IEEE 802.11 standard. At the same time, I would like to acknowledge the comments from the reviewers as well as the encouragement, patience, and help from Susanna Taggart and Michael Webb. The production staff deserves praise for producing a professional layout for the book. Finally, I am grateful to Dr. Julie Lancashire, senior commissioning editor of Artech House, who first suggested to me about writing a book and who has given me the opportunity to undertake such a meaningful project.

No one can claim to be an expert in all aspects of wireless communications and networking. Therefore, the author welcomes feedback from readers, particularly those who have discovered a more penetrating perception of the issues analyzed in this book. Comments can be sent to the author's personal e-mail address at bennybing@onebox.com.

Benny Bing
Maryland, USA

Contents

Wireless Local Area Networks

The subject area of wireless LAN has a very broad scope. It poses a variety of network engineering problems encompassing multiple protocol layers in the networking hierarchy and provides a contextual basis for cross-cutting communications and networking research. Among the key areas include network topology, frequency band allocation, data security, and the use of spread spectrum. This chapter analyzes the fundamental considerations related to wireless LAN design.

1.1   The need for wireless LANs

Traditional local area networks (LANs) link computers, file servers, printers, and other network equipment using cables. These networks enable users to communicate with each other exchanging electronic mail and accessing multi-user application programs and shared databases. To connect to an LAN, a user device must be physically connected to a fixed outlet or socket, thus creating a network of more or less stationary nodes. Moving from one location to another necessitates disconnecting from the LAN and reconnecting at the new site. Expanding the LAN implies additional cabling, which takes time to deploy, occupies more space, and increases overhead costs significantly. These factors make hard-wired LANs expensive and difficult to install, maintain, and, especially, modify.

The emergence of wireless LANs brings the benefits of user mobility and flexible network deployment in local area computing. With mobility, a network client can migrate between different physical locations within the LAN environment without losing connectivity. A more compelling advantage of wireless LANs is the flexibility to reconfigure or to add more nodes to the network without much planning effort and cost of recabling, thereby making future upgrades inexpensive and easy. The ability to cope with a dynamic LAN population generated by mobile users and portable computing devices is another major consideration for choosing a wireless LAN. Thus, the widespread use of notebook computers and handheld personal digital assistants has led to an increased dependence on wireless LANs in recent years. It has been reported in [1] that over 40 wireless LAN products are available in the market. This number is expected to rise even further with the recent introduction of the IEEE 802.11 [2] and HIPERLAN [3] wireless LAN standards.

Wireless LANs are different from wireless wide-area networks that transmit digital information over cellular or packet radio. Because these systems cover long distances, they involve costly infrastructures, provide low data rates, and require users to pay for bandwidth on a time or usage basis. In contrast, on-premise and geographically limited wireless LANs require no usage fees and provide higher data rates.

Wireless LANs deliver data rates in excess of 1 megabit per second (Mbps) and are normally used for computer to computer data transfer within a building. Being broadcast in nature, wireless LANs also allow easy implementation of broadcast and multicast services although these services must be protected from unauthorized access. In a typical wireless LAN configuration (see Figure 1.1), a transmitter/receiver (transceiver) device called an *access point* connects to the wired network from a fixed location. The access point receives, buffers, and transmits data packets between the wireless LAN and the wired network infrastructure. A single access point can support a small group of mobile nodes and can function

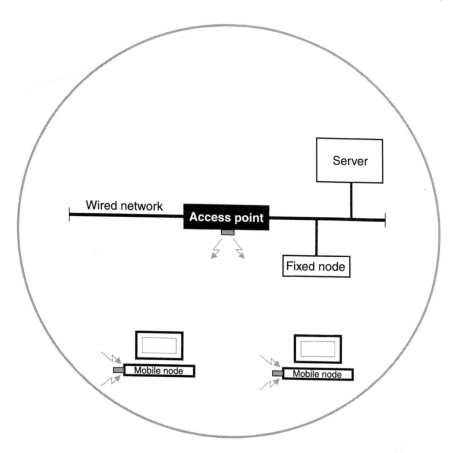

Figure 1.1 A typical wireless LAN configuration.

within a range of a few hundred meters. The antenna attached to the access point is usually mounted high but may also be placed anywhere that is practical as long as the desired radio coverage is obtained. End-user devices communicate with the access point through wireless LAN adapters which are implemented as PC cards in notebook computers, ISA or PCI cards in desktop computers, or fully integrated devices within hand-held computers (e.g., personal digital assistants, pen-based palmtop PCs) and printers. The wireless LAN adapters provide an interface between the client network operating system and the wireless link via an antenna. This enables the physical characteristics of the wireless connection to become transparent to the network operating system. Wireless LANs using portable computing devices are sometimes known as cordless LANs.

The term *cordless* emphasizes the fact that these LANs eliminate the power cord as well as the network cable.

1.2 Indoor wireless communications

Wireless LANs use either radio or infrared electromagnetic waves to transfer information from one point to another. The use of an indoor wireless link introduces new restrictions not found in conventional wired networks. The quality of the wireless link varies over space and time. Objects in a building (e.g., structures, equipment, and people) can block, reflect, and scatter transmitted signals. In addition, problems of noise and interference from both intended and unintended nodes must also be solved.

Several basic attributes of the wireless physical layer which are vastly different from the wired medium can be identified. These attributes impose fundamental limits on the range, data rate, and communications quality of wireless LANs. Wireless LANs:

- Have ill-defined network boundaries with overlaps in coverage areas;
- Suffer from limited spectral bandwidth;
- Use a shared broadcast medium;
- Lack full connectivity and are significantly less reliable than the wired physical layer;
- Have dynamic topologies with mobility functions such as roaming and handoffs adding complexity;
- Are unprotected from outside signals;
- Have time-varying propagation properties.

While wired LANs are implicitly distinct, there is no easy way to physically separate different wireless LANs. Well-defined network boundaries or coverage areas do not exist since nodes are usually mobile and transmission can occur in various locations of the network. Partial overlaps in wireless coverage areas are also common. In addition, signal propagation patterns are unpredictable and change rapidly with time. Consequently, signal coverage is not uniform, even at equal distances from the transmitter. The most prominent characteristic of wireless LANs, which differentiates them from fixed wired networks is the requirement to share a limited bandwidth among a varying number of randomly located mobile nodes.

1.3 Radio spectrum

The operating frequency band plays a significant role in wireless networking. However, it is difficult to locate radio spectrum and even harder to find spectrum that is available worldwide. Spectrum allocations are controlled by multiple international regulatory bodies (e.g., FCC in the United States, MKK in Japan, and CEPT in Europe). The usable spectrum may also vary due to frequency-selective fading and other radio propagation impairments. These two factors (i.e., lack of bandwidth and bandwidth variation) combine to produce transmission latencies and delay variation (jitter) in wireless communication.

Wireless LANs are typically designed to operate in the industrial, scientific, and medical (ISM) frequency bands because these bands are available for unlicensed operation and it is possible to build low-cost, low-power radios in this frequency range that operate at wireless LAN speeds. To prevent interference, devices operating in the ISM bands must use low-power spread spectrum transmission. The radiated power is limited to 1W in the United States, 100 mW in Europe, and 10 mW/MHz in Japan. Different frequency bands are approved for use in the United States, Japan, and Europe and any ISM wireless LAN product must meet the specific requirements for the country in which it is sold. The typical ISM frequency bands are 902 to 928 MHz GHz (26 MHz available bandwidth), 2.4000 to 2.4835 GHz (83.5 MHz available bandwidth) and 5.725 to 5.850 GHz (125 MHz available bandwidth). Since the 2.4 GHz ISM band is available in most countries (see Figure 1.2), it has been adopted by the IEEE 802.11 standard. The frequency range and available bandwidth in the 2.4 GHz band in the various countries are shown in Table 1.1. Some of these bands may have to coexist with terrestrial microwave and satellite communication frequency bands.

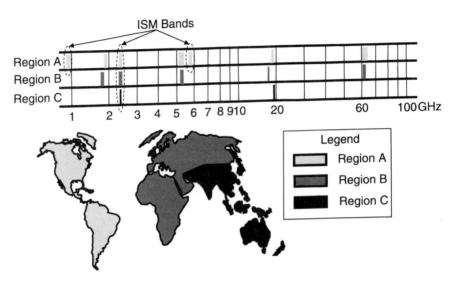

Figure 1.2 Global allocation of radio LAN spectrum.

Table 1.1
Frequency Range and Bandwidth of the 2.4 GHz ISM Bands

Country	Frequency Range (GHz)	Available Bandwidth (MHz)
United States, Most of Europe	2.400 to 2.4835	83.5
Japan	2.471 to 2.497	26
Spain	2.445 to 2.475	30
France	2.4465 to 2.4835	37

1.4 Path loss

The transmitted signal power normally radiates (spreads out) in all directions and attenuates quickly with distance. Thus, very little signal energy reaches the receiver, giving rise to a proportional relationship between distance and path loss (see Figure 1.3). The simplest model for path loss is an exponential relationship. In free-space, the path loss exponent is 2. However, the signal attenuation is dependent not only on distance and transmitted power but also on reflecting objects, physical

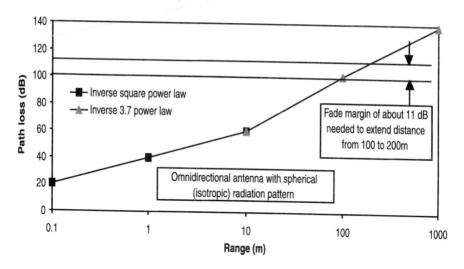

Figure 1.3 Path loss in a typical 2.4 GHz indoor wireless link.

obstructions, and the amount of mutual interference from other transmitting nodes. Small changes in position or direction of the antenna, shadowing caused by blocked signals, and moving obstacles (e.g., people and doors) in the environment may also lead to drastic fluctuations in signal strength. Similar effects occur regardless of whether a node is stationary or mobile. Hence, while the free-space exponent may be relevant for short distance transmission (e.g., up to 10 m), the path loss is usually modeled with a higher-valued exponent of 3 to 5 for longer distances.

The wide variation in path attenuation must be continually adapted by wireless systems. A fade margin can help to offset the changes in signal attenuation. The amount of fade margin is dictated by antenna gain constraints and the maximum transmitted power specified by regulatory authorities.

1.5 Multiple access

For radio LANs, sharing of bandwidth is essential since radio spectrum is not only expensive but also inherently limited. This is in contrast to wired networks where bandwidth can be increased arbitrarily by adding extra cables. However, the broadcast nature of the wireless link poses a difficult problem for multiple access in that the success of a transmission is no longer independent of other transmission. To make a transmission successful, interference must be avoided or at least controlled. Otherwise, multiple transmissions may lead to collisions and corrupted signals. A multiple access (or medium access control) protocol is required to resolve these access contentions among nodes and transform a broadcast wireless network into a logical point-to-point network.

In general, multiple access protocols can be categorized under contention, reservation, polling, and static allocation (see Table 1.2). At one extreme, where no control is enforced, two or more nodes may transmit at the same time and conflicts may occur. These uncontrolled (contention) schemes are very easy to implement but pay the price in the form of wasted bandwidth due to collisions [4]. The other extreme is represented by a rigid system of fixed control (static allocation) where each node is permanently assigned a portion of the total bandwidth for its exclusive use. Such control places hard limits on the number of users sharing a given bandwidth. Examples include frequency division multiple access (FDMA) and time division multiple access (TDMA). The class of dynamic control protocols (reservation and polling) dedicates a small part of the bandwidth for control and this control information is used to determine the identity of the nodes with data to transmit. While the performance of contention and reservation techniques is dependent on the combined traffic from all nodes in the network, the performance of fixed allocation and polling schemes is strongly influenced by the traffic requirements of each individual node.

The multiple access capability of spread spectrum systems is distinctly different from narrowband systems. Not only can simultaneous transmission (from nodes with data to transmit) be tolerated, the number of such a transmission should be large in order to achieve high network

Table 1.2
Comparison of Multiple Access Protocols

Multiple Access Protocol	Collisions	Control Overhead	Idle Time
Contention	Yes	No	No
Reservation	No	Yes	No
Polling	No	Yes	No
Fixed Allocation	No	No	Yes

capacity. Thus, as long as different receivers are involved, multiple access protocols in spread spectrum systems are designed to have multiple transmissions taking place at the same time. In this case, interference may sometimes dominate over noise as an error-producing mechanism. Spread spectrum multiple access gives the advantage of soft capacity where no absolute limit is placed on the number of users. Performance is degraded in proportion to any increase in users on the system. However, spread spectrum multiple access has not been implemented in wireless LANs. Instead, most wireless LANs use some form of carrier sense multiple access (CSMA) which is a contention-based multiple access protocol. Details of this protocol are provided in Chapter 4.

1.6 Multipath

Multipath is a critical problem that needs to be dealt with as it produces a variable bit error rate that can lead to intermittent network connectivity. Multipath occurs even when a single node is transmitting whereas interference resulting from multiple access is due to transmissions from two or more nodes. The multipath phenomenon is caused by differently delayed versions of the original signal and this leads to a superposition of multiple signals with a delay spread at the receiver (see Figure 1.4). The delayed signals are produced by reflected signals arriving at the

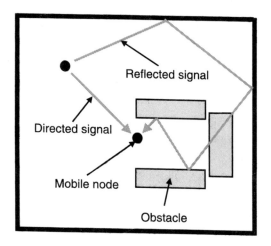

Figure 1.4 Multipath propagation.

receiver along different paths, thus resulting in differing propagation delays. Reflections are generated when the propagating signal impinges on an object which is large compared to its wavelength (e.g., walls). Scattering occurs when objects smaller than the wavelength of the propagating signal are encountered. Since these delayed signals are dispersed in time, they may arrive at the receiver with different phase shifts and this causes the received signal (which is usually modulated on a high-frequency carrier) to become stronger or weaker. The summation of many such randomly phased components results in a signal amplitude that is Rayleigh distributed and a function of location. Most wired and infrared LANs use baseband or unmodulated transmission and these LANs do not suffer such problems since the phase component of the received signal is not required. Multipath may also lead to frequency-dependent fading that causes a more destructive effect called intersymbol interference. In this case, the delay spread approaches the duration of a data symbol, making the symbols overlap in varying degrees at the receiver. A data symbol refers to an analog signal waveform that is used to represent one or more data bits. Generally, the severity of the multipath problem is determined by the number of reflective surfaces in the environment and the distance between the transmitter and receiver. The effects of multipath are most noticeable at the fringe area of reception.

1.6.1 Delay spread

The delay spread is defined as the difference in propagation delay between the directed (line-of-sight) signal and the reflected signal that takes the longest path from transmitter to receiver. Because the delay spread is a random variable, it is often represented by its standard deviation (called the root-mean square or rms). The delay spread can be used to characterize different types of signal fading. If the product of the rms delay spread and the signal bandwidth is much less than 1, then the fading is called flat fading. If the product is greater than 1, then the fading is classified as frequency-selective.

1.6.2 Flat fading

In flat fading, the received signal envelope has a random signal amplitude and phase that usually follows the Rayleigh probability density function.

To illustrate how flat fading can distort the amplitude and phase of a received signal, consider a sinusoidal signal directly transmitted to the receiver and the same signal being reflected and then received. For simplicity, it is assumed that the received signal comprises the sum of the directed signal and the reflected signals (each with a random amplitude and phase). Whether the sum of two such modulated signals cancel or reinforce each other strongly depends on the difference in their phase angles. If this phase difference is near 180 degrees, then the net result is a deep fade in the received signal. As can be seen in Figure 1.5(c), these fades are separated by about half a wavelength, which at 2.4 GHz is approximately 0.06 m. If a mobile node moves at 10 km/hr (pedestrian speed) in this frequency band, fades can be expected every 22 ms. Clearly, the rate of fading is proportional to the velocity of the user motion.

The phase of the received signal may also differ considerably from the directed signal (see Figures 1.5(a–f)). If the reflected signal is attenuated, the impact on the amplitude and phase of the received signal becomes limited (compare Figures 1.5(b) and 1.5(d)). A reflected signal need not always produce a negative effect in a multipath link. As shown in Figure 1.5(e), although reflected signal 1 virtually cancels out the directed signal, reflected signal 2 actually provides a means to recover the original signal. This diversity property is inherent in multipath propagation and is some-

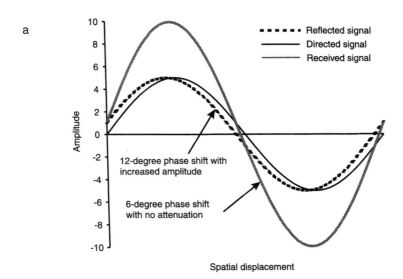

Figure 1.5 Impact of multipath reflections on received signal.

b

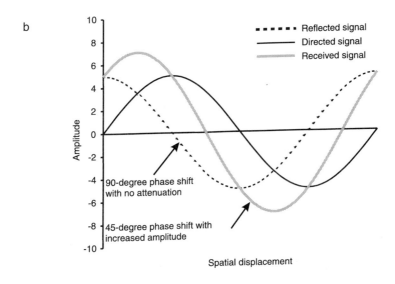

c

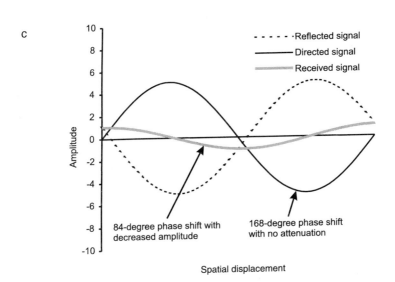

Figure 1.5   (continued).

d

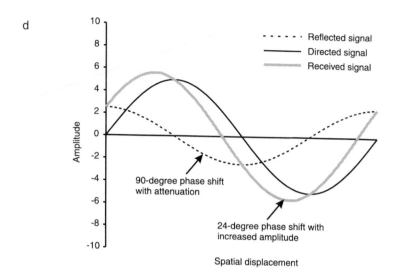

e

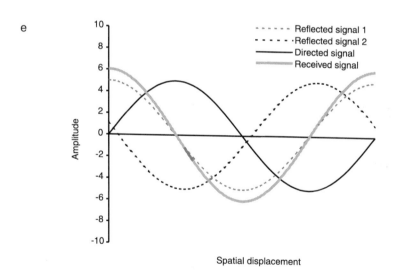

Figure 1.5 (continued).

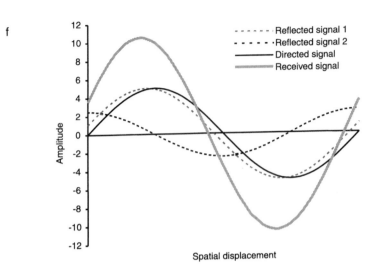

Figure 1.5 (continued).

times exploited by radio receivers to improve reception. It is interesting to note that in all cases, the frequency of the received signal remains unaffected by the phase or amplitude distortions of the reflected signals.

1.6.3 Frequency-selective fading

Multipath often leads to frequency-selective fading which is nonuniform fading over the frequency band occupied by the transmitted signal (see Figure 1.6). The fades (notches) are usually correlated at adjacent frequencies and are decorrelated after a few MHz. The severity of such fading depends on how rapidly the fading occurs relative to the round-trip propagation time on the wireless link. It is clear that while being in a fade, no reasonable amount of error-correction coding can help to mitigate the problem. On the other hand, Figure 1.6 shows that most of the frequency band is free from severe fading at any given time.

The coherence bandwidth refers to the range of frequencies over which signal fading is correlated. The fades occur in the received signal spectrum spaced C Hz apart where C is the coherence bandwidth. If the signal components arrive with a differential rms delay spread of D seconds, C is in the order of $1/D$ Hz. If the coherence bandwidth is small compared to the bandwidth of the transmitted signal, then the wireless link is frequency-selective and different frequency components are sub-

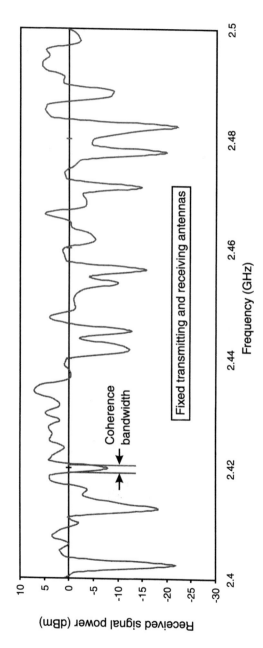

Figure 1.6 Frequency-selective multipath fading in a typical 2.4 GHz indoor wireless link.

ject to different amplitude gains and phase shifts. Conversely, a wireless link is nonfrequency-selective if all frequency components are subject to the same attenuation and phase shift. Frequency-selective fading poses a more serious problem since matched filters that are structured to match the undistorted part of the spectrum will suffer a loss in detection performance when the attenuated portion of the spectrum is encountered. Either the data rate must be restricted so that the signal bandwidth falls within the coherence bandwidth of the link or other techniques such as spread spectrum must be used to suppress the distortion. The rms delay spread in an indoor environment can vary significantly from 30 ns in a small room to 300 ns in a factory or warehouse. This corresponds to a coherence bandwidth of roughly between 3 MHz and 33 MHz.

1.6.4 Intersymbol interference

In general, the effect of the delay spread is to cause the smearing of individual symbols in the case where the symbol rate is sufficiently low, or to further cause time-dispersive fading and intersymbol interference if the symbol rate is high (see Figure 1.7). Intersymbol interference is a form

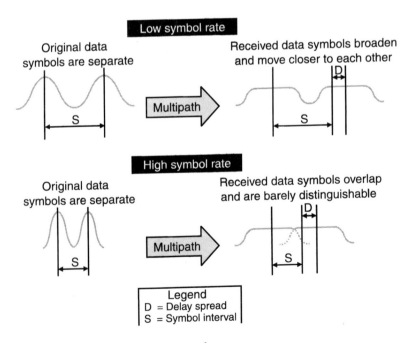

Figure 1.7 Effects of delay spread.

of self-interference that increases the error rate in digital transmission, an impairment that cannot be overcome simply by improving the signal-to-noise ratio. This is because increasing the signal power in turn increases the self-interference. For a data rate of 2 Mbps with 2 bits/symbol, the symbol rate is 1 Msymbol/s. A rms delay spread of 100 ns will cause adjacent symbols to overlap by 1 Msymbol/s × 100 ns or 0.1 symbol. At higher symbol rates or larger delay spreads, the difference in delay among the various signal reflections arriving at the receiver can be a significant fraction of the symbol interval. Normally, a delay spread of more than half a symbol interval results in indistinguishable symbols and a sharp rise in the error rate.

1.6.5 Countermeasures against multipath

Several techniques to mitigate the effects of multipath are commonly employed by wireless LAN systems. Since the effects of multipath changes with distance and frequency, such interference can be minimized by altering antenna position or changing operating frequency. In antenna diversity, the best signal from multiple reflections are selected among two or more antennas. Two antennas separated by about an odd multiple of a quarter of a wavelength is enough to cause almost independent fades at the receiving antennas (see Figure 1.8). At 2.4 GHz, a quarter wavelength is roughly equivalent to 3 cm. Maximal ratio combining can be employed where signals received by different antennas are weighted and combined to produce the best signal-to-noise ratio [5]. Another method involves selection diversity which requires the signal strength at each antenna to be monitored by an individual receiver. In switched diversity, a single receiver switches from one antenna to another when the antenna is in a fade. Alternatively, different antenna polarization can be employed to keep the antenna dimensions small. These diversity techniques are effective in overcoming multipath fading without any bandwidth penalty.

Spread spectrum is another solution to deal with multipath interference. Spread spectrum is a powerful combination of bandwidth expansion (beyond the coherence bandwidth) and coding, whereby the former combats intersymbol interference and the latter allows individual symbols smeared by multipath to be received correctly.

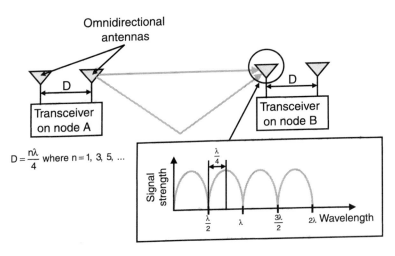

Figure 1.8   Antenna diversity to combat the multipath problem.

Yet another method involves multilevel (as opposed to binary) modulation where each symbol is mapped into multiple data bits, thus increasing the data rate while keeping the symbol rate sufficiently low. For example, a 2 Mbps bit stream using differential quadrature phase shift keying (DQPSK) modulation results in a symbol rate of 1 Msymbol/s. Unfortunately, multilevel modulation demands a high signal-to-noise ratio for good performance and may not be appropriate for noisy or power limited wireless LANs, especially infrared LANs.

Equalization may also be used to combat multipath. The simplest equalization technique (also known as linear equalization) amplifies the attenuated part and attenuates the amplified part of the spectrum. Such a technique attempts to invert and neutralize the effects of the medium. This can be achieved by passing the received signal through a filter with a frequency response that is the inverse of the frequency response of the medium. When additional noise, interference, or frequency-selective fades are present at the input to the equalizer, the equalizer must minimize distortion without adversely enhancing noise. In this case, more sophisticated equalizers are needed. The decision-feedback equalizer (DFE) is a nonlinear equalizer that does not enhance the noise because it estimates the frequency response of the link rather than inverting it. However, DFEs may result in error propagation and are useful only when the medium exhibits reasonably low bit error rate without coding. When applied to slowly varying indoor wireless links, equalization requires minimum overhead, typically 14 to 30 symbols for every 150 to 200 data symbols. For rapid variations in the wireless link, equalization can be difficult because these problems have to be alleviated using a long equalizer training sequence. Nevertheless, by using novel joint data and link estimation algorithms, data symbols can still be reliably recovered even after the link undergoes fading [6].

Although the use of these techniques can help lessen the impact of multipath problems such as fading and interference, traditional error control methods based on retransmission or error correction coding may still be required to mitigate the errors introduced by the multipath link. However, error-correction coding is not very effective in typically slow-varying indoor channels (see Section 6.3.12 for a fuller discussion on the application of error-correction coding).

1.7 Shadow fading

Besides fading due to multipath propagation, physical obstructions can cause shadow fading where the transmitted signal power is blocked and hence severely attenuated by the obstruction. The variation in received signal power due to these obstructions is called shadow fading and depends on the number and dielectric properties of the obstructing objects. Measurement studies have shown that signal power variation under shadow fading is spectrally flat and exhibits a log-normal or Gaussian distribution. Increasing the transmit power can help to mitigate the effects of shadow fading although this reduces battery life and can cause interference for other users.

1.8 Propagation delay

One key advantage of wireless local area communications is the short round-trip propagation delay that permits the rapid transmission of control information (e.g., polling signals and acknowledgments) and significantly reduces guard time overheads between successive packet transmission. The ratio between the propagation delay and the packet transmission time is an important parameter since the performance of many multiple access schemes such as CSMA degrades rapidly when the packet transmission time becomes small compared to the propagation delay.

Consider a typical wireless LAN that spans 30 m. Using a link propagation speed of 3×10^8 m/s, the propagation delay becomes 100 ns. The average multipath delay is in the region of 100 ns which restricts the maximum data rate to about 1/(100 ns) or 10 Mbps. This estimated data rate should be regarded as the low end of the range of data rates achievable with wireless communications. Data rates can be increased considerably by using diversity reception and sectored antennas, employing multilevel signaling and coding techniques, implementing adaptive equalization at the receiver, and adopting wideband transmission such as spread spectrum. If the average packet length is 10,000 bits, then for a data rate of 10 Mbps, the average packet transmission time becomes 1 ms. The ratio of the propagation delay to the average packet transmission time yields 0.0001, a value small enough for CSMA to operate efficiently. In this case, the packet transmission time will only approach the propagation delay when the data rate approaches 10,000 bit/100 ns or 100 Gbps. Such high data rates are unlikely to be achievable in wireless LANs. Note that for practical implementation, parameters such as carrier sensing time, processing delay, and transceiver turnaround time have to be considered in addition to the propagation delay.

1.9 Summary

Mobility and flexibility make wireless LANs effective extensions and attractive alternatives to wired networks. Wireless LANs provide all the functionality of wired LANs, but without the physical constraints of the wire itself. However, the wireless link has some unique obstacles that need to be solved. The medium is a scarce resource that must be shared among network nodes. It can be noisy and unreliable where the transmission from mobile nodes interferes with each other to varying degrees. The transmitted signal power dissipates rapidly in space and becomes attenuated. Physical obstructions may block or generate multiple copies of the transmitted signal. The received signal strength normally changes slowly with time because of path loss, more quickly with shadow fading and very quickly because of multipath flat fading.

References

[1]   "1997 Global Enterprise Networking Directory," *Data Communications International*, Vol. 25, No. 11, August 1996.

[2]   IEEE P802.11, *Information Technology–Telecommunications and Information Exchange Between Systems– Local and Metropolitan Area Networks–Specific Requirements, Part 11: Wireless LAN Medium Access Control (MAC) and Physical Layer (PHY) Specifications*, November 1997.

[3]   EN 300 652, "Broadband Radio Access Networks (BRAN); High Performance Radio Local Area Network (HIPERLAN) Type 1; Functional Specification," October 1998.

[4]   Kleinrock, L. "On Queueing Problems in Random Access Communications," *IEEE Transactions on Information Theory*, Vol. IT–31, No. 2, March 1985, pp. 166–175.

[5]   Acampora, A. "System Applications for Wireless Indoor Communications," *IEEE Communications Magazine*, Vol. 25, No. 8, August 1987, pp. 11–20.

[6]   Seshadri, N., C. Sundberg and V. Weerackody, "Advanced Techniques for Modulation, Error Correction, Channel Equalization and Diversity," *AT&T Technical Journal*, Vol. 72, No. 4, July-August 1993, pp. 48–63.

Selected Bibliography

Ahmadi, H., A. Krishna and R. LaMaire., "Design Issues in Wireless LANs," *Journal of High Speed Networks*, Vol. 5, 1996, pp. 87–104.

Chevillat, R. and H. Rudin, "Wireless Indoor Communication," *IEEE Network*, November 1991, Vol. 5, No. 6, p. 10.

The Evolution of Untethered Communications, Washington, D. C.: National Academy Press, 1997.

Kahn R., S. Gronemeyer, J. Burchfiel and R. Kunzelman, "Advances in Packet Radio Technology," *Proceedings of the IEEE*, Vol. 66, No. 11, November 1978, pp. 1468–1496.

Links, C., W. Diepstraten and V. Hayes, "Universal Wireless LANs," *Byte*, May 1994, pp. 99–104.

National Science Foundation, Division of Networking and Communications Research and Infrastructure, "Research Priorities in Wireless and Mobile Communications and Networking," May 1994.

National Science Foundation, Division of Networking and Communications Research and Infrastructure, "Research Priorities in Wireless and Mobile Communications and Networking," March 1997.

Steenstrup, M. "Mobile Communications," Guest Editorial, *IEEE Network*, March/April 1994, Vol. 8, No. 2, p. 5.

Wickergren, I. "Local-Area Networks Go Wireless," *IEEE Spectrum*, September 1996, pp. 34–40.

The Wireless LAN Alliance, "Introduction to Wireless LANs," 1996.

Classification of Wireless LANs

Wireless LANs can be broadly classified under radio or infrared LANs. Radio LANs can be based on narrowband or spread spectrum transmission while infrared LANs can be diffuse or directed. This chapter surveys the major types of radio and infrared LANs and evaluates the strengths and weaknesses of each category.

2.1   Radio LANs

Many wireless LAN systems use spread spectrum technology. The general concept of spread spectrum has been in existence for more than 50 years and was originally developed by the military for reliable and secure communications. Spread spectrum refers to signaling schemes which are based on some form of coding (that is independent of the transmitted information) and which use a much wider bandwidth beyond what is required to transmit the information. The wider bandwidth means that interference and multipath fading typically affect only a small portion of a spread spectrum transmission. The received signal energy is therefore relatively constant over time. This in turn produces a signal that is easier to detect provided the receiver is synchronized to the parameters of the spread spectrum signal. Spread spectrum signals are able to resist signal interference and are hard to detect and intercept. There are two spread spectrum techniques: direct-sequence spread spectrum (DSSS) and frequency-hopping spread spectrum (FHSS). The operation of these techniques is illustrated in Figure 2.1.

2.2 Direct-sequence spread spectrum

DSSS systems offer highly reliable transmission with relatively small signal-to-noise ratios. DSSS spreads the energy (power) of the signal over a large bandwidth. The energy per unit frequency is correspondingly reduced. Hence, the interference produced by DSSS systems is signifi-cantly lower compared to narrowband systems. This allows multiple DSSS

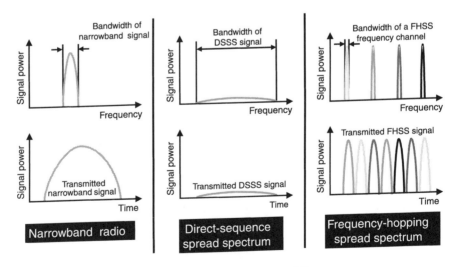

Figure 2.1 Narrowband radio, DSSS, and FHSS.

signals to share the same frequency band. To an unintended receiver, DSSS signals appear as low-power wideband noise and are rejected by most narrowband receivers. Conversely, this technique diminishes the effect of narrowband interference sources by spreading them at the receiver.

DSSS combines the data stream with a higher speed digital code. Each data bit is mapped into a common pattern of bits known only to the transmitter and the intended receiver. The bit pattern is called a pseudonoise code and each bit in the code is called a chip. The term chip is used to emphasize the fact that one bit in the pseudonoise code form part of the actual data bit. The sequence of chips within a bit period is random but the same sequence is repeated in every bit period, thus making it pseudorandom or partially random. The chipping rate of an n-bit pseudonoise code is n times higher than the data rate. It is the high rate for the chipping sequence that results in a very wide bandwidth. Figure 2.2 shows how information bits can be spread out 11 times through the use of an 11-chip pseudonoise code. This 11-chip code is employed by the IEEE 802.11 standard (see Section 4.3.2). The chip rate is 11 times faster than the data rate. The longer the pseudonoise code, the greater the probability that the original data can be recovered, but more bandwidth is required since a higher chipping rate is required. Usually, the spreading bandwidth is about twice the chip rate. Thus, a chip rate of 11 Mchip/s translates to a spreading bandwidth of 22 MHz. Note that any

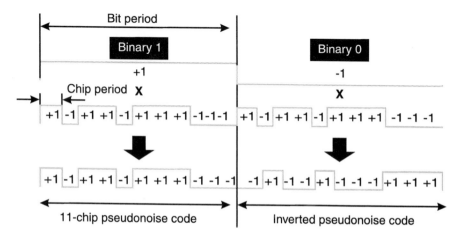

Figure 2.2   DSSS using an 11-chip pseudonoise code.

of the 11 chips in the pseudonoise code in Figure 2.2 can be selected as the first chip (i.e., the starting chip). Each of these starting chips corresponds to a different phase of the same code.

At the receiver, the chips are despread by the same pseudonoise code and mapped back into the original data bit. However, the energy of noise and interference that may have been added during the transmission are spread and hence, suppressed, by the pseudonoise code (see Figure 2.3). In addition to knowing the pseudonoise code used by the transmitter, the receiver must also be properly synchronized to the correct phase of the code as well as its chip rate. Thus, a challenging function of the timing mechanism in the preamble of a DSSS packet is to enable the receiver to synchronize to the correct phase of the pseudonoise code of a packet in the shortest possible time. Since the transmission of packets is asynchronous, every DSSS packet must be preceded by a preamble for synchronization purposes.

When the pseudonoise code generated by the receiver is exactly synchronized to the received signal, the despreading process produces peaks of high autocorrelation (see Figure 2.4). If the pseudonoise code is shifted by one or more chip intervals, low autocorrelation results (see Figure 2.5). Similarly, noise and interference that may have been added into the received signal produce low autocorrelation since they are usually not correlated to the pseudonoise code. This gives rise to a single peak within each bit interval (see Figure 2.6). When transmitted over a wireless link with very little delay spread, the sidelobes of the received signal will either add constructively or destructively but the peak will not be affected. Because the autocorrelation peaks occur periodically, DSSS receivers can simply lock on to these peaks when decoding data after they have acquired the initial synchronization. This implies that even if one or more chips are damaged during transmission, the autocorrelation property embedded in the DSSS signal can recover the original data without the need for retransmission. In addition, interfering peaks from multipath reflections (see Figure 2.7) or new packet transmission can be rejected provided these peaks do not coincide with the desired peaks. The ability to tolerate interfering peaks is dependent on the time resolution of the DSSS signal. A higher resolution requires wider spreading but allows interfering peaks to be resolved easily. Note that a DSSS receiver synchronizes to one of the multipath reflections received. Other reflections delayed by more than one chip duration are significantly attenuated.

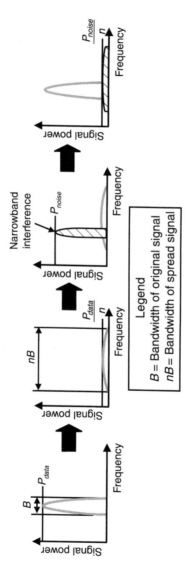

Figure 2.3 Narrowband interference rejection in DSSS.

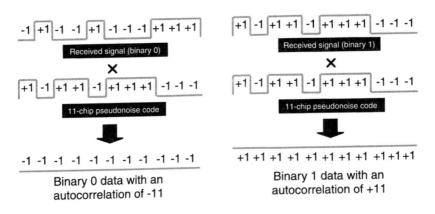

Figure 2.4 Synchronized pseudonoise codes produce high autocorrelation.

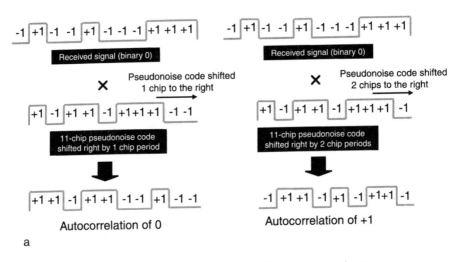

Figure 2.5 Time shifted pseudonoise codes produce low autocorrelation.

A key parameter for DSSS systems is the number of chips per bit which is called the processing gain or the spreading ratio. A high processing gain increases the ability of the signal to reject interference (since the interference is spread by a factor equivalent to the processing gain). A low processing gain increases the amount of bandwidth available for data transmission. In the United States, the FCC specifies a minimum process-

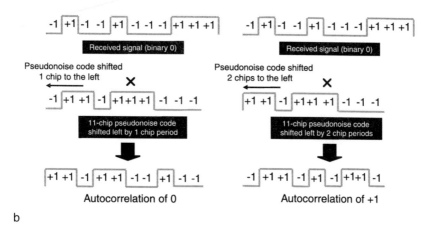

b

Figure 2.5 (continued).

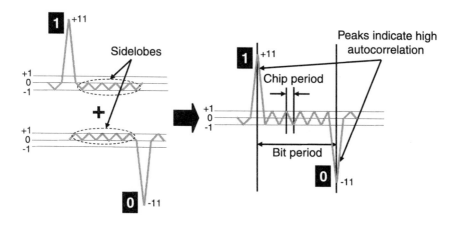

Figure 2.6 Addition of autocorrelation peaks in DSSS signals.

ing gain of 10 for the ISM bands. The upper limit of the processing gain is determined by the available bandwidth. The IEEE 802.11 standard uses an 11-chip pseudonoise code that spreads the data 11 times before transmission, thus providing 10.4 dB of processing gain. This gain is quite small compared to spread spectrum cellular systems. Hence, interference suppression is limited but more bandwidth is available for high-speed transmission of user data.

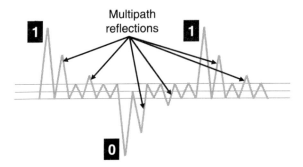

Figure 2.7 Autocorrelation peaks due to multipath reflections.

The DSSS method used in wireless LANs differs from code division multiple access (CDMA). Code division refers to the fact that transmission with orthogonal pseudonoise codes may overlap in time with little or no effect on each other. Different nodes transmit using unique codes. For a receiver tuned to the code of one transmission, other signals (using other codes) appear as background noise. In the despreading process, this noise will be suppressed by the processing gain. A random access CDMA system requires complex receivers that can synchronize and demodulate all pseudonoise codes. DSSS wireless LANs utilize the same pseudonoise code and therefore do not have a set of codes available as is required for CDMA operation [1]. A single code enables information to be broadcast easily. In addition, the code can be made shorter, thereby increasing the bandwidth for data transmission.

2.3 Frequency-hopping spread spectrum

FHSS spreads the signal by transmitting a short burst on one frequency channel and then changes (hops) to another channel for another short period of time in a predefined pattern known to both transmitter and receiver. Unlike DSSS which uses multiple frequency channels simultaneously, FHSS uses multiple frequency channels randomly. Because the frequency channels are narrowband, they provide excellent signal-to-noise ratio and narrow filters can be employed to reject interference. To an unintended receiver, FHSS transmission appears to be short-duration impulse noise. The hopping pattern determines the selected frequency channels and the order in which the channels are used. Synchronization between transmitter and receiver must be acquired and maintained so that they are hopping on the same frequency channel at the same time. For FHSS systems, the processing gain is defined as the ratio of the total bandwidth occupied by the frequency channels over the signal bandwidth.

The dwell time (i.e., the time at each frequency channel) must always be specified since the nature of FHSS requires channels to be changed after a certain time interval. The number of frequency channels in a hopping pattern and the dwell time are restricted by most regulatory agencies. For example, in the 2.4 GHz band, the FCC requires that 75 or more channels be used in each pattern and that the maximum dwell time be 400 ms (out of 30 s). To ensure that the available channels are equally chosen on the average, all channels within a hopping pattern must be used up before the channels in the pattern are reused, thus giving a minimum hopping rate of 75/30 or 2.5 hop/s. This means that most of the time, many data packets can be transmitted within the dwell time of a single frequency channel in a hopping pattern. Consider the transmission of a maximum-length Ethernet data packet of 1,518 octets or 12,144 bits. With a 2 Mbps data rate, the transmission time for the packet is roughly 6 ms. Hence, more than 60 maximum-length Ethernet frames can theoretically be sent within the dwell time of 400 ms. Since normal applications use much shorter packets, this number increases dramatically.

The ratio between the hopping rate and the data rate results in two modes of FHSS. When the hopping rate is higher than the data rate, the system is known as fast frequency hopping. Conversely, when the hopping rate is lower (as in the example above), the system is classified as slow frequency hopping. The hopping rate has a profound effect on the performance of a FHSS system. Unlike licensed narrowband systems that operate in dedicated spectrum and are generally not concerned with interference, frequency hopping systems operating in shared and unlicensed ISM bands may experience interference on some channels. For slow FHSS systems, this can potentially lead to the loss of many data packets. Thus, fast frequency hopping usually outperform slow hopping, even with the same processing gain. However, fast FHSSs are expensive to implement since they require very fast frequency synthesizers.

Hopping patterns are designed to be almost orthogonal so that frequency channels in different patterns hardly interfere with one another. Thus, the collision probability among channels is small—typically about 1%. Successive hops (between different frequency channels) usually exceed the coherence bandwidth of the link. Hence, if interference interrupts data transmission on a particular hop, there is only a slight chance that it will affect the next hop in the pattern. For fast-hopping FHSSs, such collisions typically result in random errors that can be corrected by the receiver without retransmission (see Figure 2.8). However, given the slow hopping rate in FHSS wireless LANs, collisions are likely to lead to one or more corrupted packets that require retransmission to restore the lost data. Unlike DSSS wireless LANs which employ a common pseudonoise code, FHSS wireless LANs may employ more than one hopping pattern to increase network capacity. Figure 2.9 illustrates how different hopping patterns can be assigned to different coverage areas serviced by different access points. The idea of using multiple hopping patterns is equivalent to using different pseudonoise codes.

2.4 Comparison of DSSS and FHSS wireless LANs

Since DSSS systems are interference averaging whereas FHSS systems are based on interference avoidance, each system has its individual strengths and weaknesses.

2.4.1 Overall network capacity

If single frequency channels are compared, DSSSs have the potential for greater transmission speeds since DSSS channels are typically wider than FHSS channels. For example, in the 2.4 GHz ISM band, each DSSS channel occupies a bandwidth of about 22 MHz whereas with FHSS, the maximum bandwidth is specified at 1 MHz (see Figure 2.10). Although bandwidth spreading is required in DSSS wireless LANs, such systems are already providing wireless data rates of up to 11 Mbps per channel whereas the highest FHSS data rate currently stands at 3 Mbps per channel.

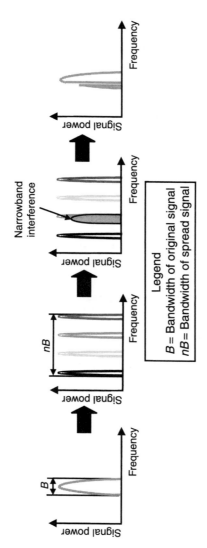

Figure 2.8 Narrowband interference rejection in FHSS.

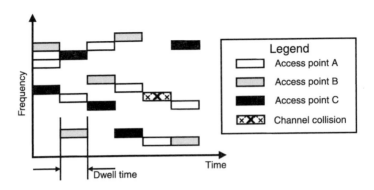

Figure 2.9   Overlapping FHSS transmission using different hopping patterns.

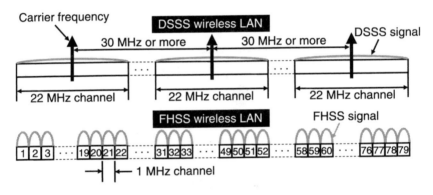

Figure 2.10   Frequency channels for 2.4 GHz DSSS and FHSS wireless LANs.

DSSS systems, however, are not as scalable as FHSS systems. The wider bandwidth allocated to each DSSS channel becomes a liability since less nonoverlapping channels are available. This limits the number of independent wireless coverage areas that can colocate and operate without interference. In the 2.4 GHz ISM band, the maximum number of colocated DSSS coverage areas is 3. On the other hand, due to a higher number of nonoverlapping 1 MHz channels and consequently, more hopping patterns, 2.4 GHz FHSS wireless LANs can potentially support up to 26 colocated coverage areas, thus delivering higher aggregate network capacity. However, this option is costly since more access points are required.

2.4.2 Interference rejection

A key difference between 2.4 GHz FHSS and DSSS wireless LANs is that in
FHSS, the frequency channels in a hopping pattern are spread across the
entire ISM band, whereas for DSSS, only a portion of the ISM bandwidth
is used. Hence, FHSS wireless LANs are less susceptible to interference
that may happen to occupy a fixed portion of the 2.4 GHz ISM band. This is
illustrated in Figure 2.11 which assumes that five frequency channels are
selected for a hopping pattern and that each channel has a dwell time of
400 ms. Clearly, time-invariant and wideband interference sources can
degrade the performance a DSSS system more seriously than a frequency
agile FHSS system. Moreover, higher power FHSS transmission can
overcome and reduce the effects of lower power interference. DSSS
systems cannot avoid the interference and their low power transmission
typically cannot overpower interference. However, because the duration
of the dwell time is relatively long, several FHSS packets may still be lost
due to interference.

DSSS can show some improvement if antenna diversity is used so that
the transmitted signal is received in different locations. In addition, when
two or more multipath signals are separated in time by more than one

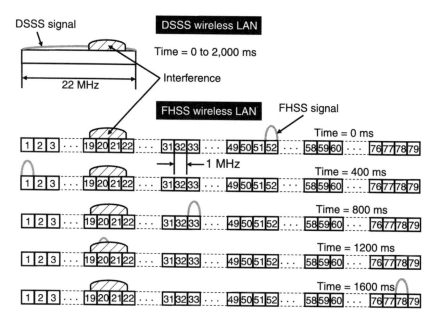

Figure 2.11 Interference rejection in DSSS and FHSS wireless LANs.

chip duration, they become independent of one another. These independent multipaths can be combined to give diversity at the DSSS receiver, a process that actually enhances the performance of the receiver.

2.4.3 Collisions

A collision in a spread spectrum system has about the same effect as a collision in a narrowband system. Packets involved in the collision are discarded and must be retransmitted at a later time. An important difference that arises in a spread spectrum system is the possibility that even when two packets collide, a third packet is received (e.g., the third packet may be using a time-shifted pseudonoise code or hopping pattern). Thus, if spread spectrum transmission is employed, collision events are associated with individual packets rather than with specific time slots or with individual receivers [2].

In general, collisions in both DSSS and FHSS systems occur much less frequently than narrowband systems since it is highly unlikely that asynchronous packet transmission will coincide exactly in time. Thus, when different wireless coverage areas with the same frequency channels overlap, multiple spread spectrum transmission may lead to random errors but not necessarily destructive collisions. For instance, DSSS receivers are capable of capturing the first arriving data packet even when there are subsequent, time-overlapping packet transmission adopting the same pseudonoise code. This is because late arriving packets produce different time-shifts (phases) of the same code, whose autocorrelation peaks are then offset in time. Suppose packets A and B are transmitted to node C using DSSS. Packet B is delayed with respect to packet A by more than one chip period (see Figure 2.12). In a narrowband system, any 2 packets that are overlapped in time results in the destruction of both. The vulnerable collision duration is the entire length of a data packet. With spread spectrum, the vulnerable period is considerably shorter, typically within the preamble bits of the data packet. Thus, if node C successfully synchronizes to the autocorrelation peaks of packet A's preamble, it will have little difficulty tracking other peaks belonging to the rest of the data packet since these peaks occur periodically. Interfering peaks from packet B can simply be ignored. Hence, packet B can be considered lost but there is no collision. An implicit assumption in this example is that the autocorrelation peaks of packets A and B do not coincide. It is therefore

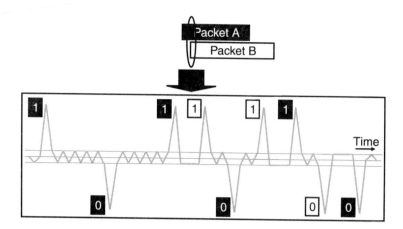

Figure 2.12 Overlapping DSSS transmission using a common pseudonoise code.

desirable that the pseudonoise code be long enough to cover the entire length of packet A so that it is guaranteed to be collision-free once a receiver has synchronized to its preamble. If the code is repeated within the lifetime of packet A, packet B may still destroy packet A if the autocorrelation peaks overlap. Note that if packets A and B are addressed to different nodes and the autocorrelation peaks from these packets do not overlap, it is possible to decode both packets without error.

Like DSSS systems, FHSS allows multiple transmission using delayed versions of the same hopping pattern (Figure 2.13). The transmission will then be combined in a way similar to frequency division multiplexing. Collisions will only occur if two or more nodes adopt the same time-shifted hopping pattern.

2.4.4 Carrier sensing

Almost all wireless LANs employ some form of carrier (or activity) sensing where transmission is deferred when an ongoing packet transmission is detected. For spread spectrum networks, the goal of carrier sensing is to determine whether the intended receiver is busy rather than to determine if there are other packets transmitted within the range of the receiver. This requires the address of a packet to be decoded in order to identify the intended recipient. Unfortunately, spread spectrum wireless LANs do not decode the address of the packet when performing carrier sensing.

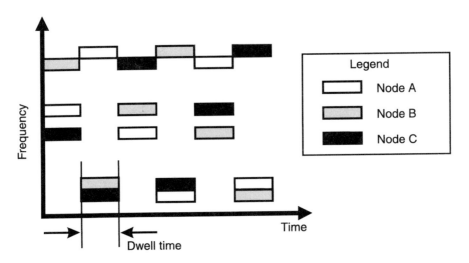

Figure 2.13 Overlapping FHSS transmission using a common hopping pattern.

For instance, in DSSS systems, carrier sensing is normally implemented by monitoring the interference level or by checking the presence of autocorrelation peaks of a common pseudonoise code. Carrier sensing conducted in this manner virtually eliminates overlapped (but non-interfering) packet transmissions. Although it may appear that carrier sensing in spread spectrum systems serves no useful purpose if the state of the intended recipient is unknown, it does allow a node to transmit at a higher data rate. If multiple autocorrelation peaks are combined within the same transmission, the positions of these peaks can then be used to represent more information bits with no increase in the bandwidth requirement. For an 11-bit pseudonoise code, up to four time-shifted peaks can operate simultaneously, thus providing a four-fold increase in data rate for a single node with some enhancement of the sidelobes [3]. Note that a short pseudonoise code used repeatedly within a packet transmission allows carrier sensing in DSSS systems to be performed quickly and at any point in the packet.

2.4.5 Health concerns

Extensive scientific research provided no conclusive evidence that radio transmission has an adverse biological effect on the health of users.

Moreover, the radiation emitted by wireless LANs is likely to be more benign than mobile cellular phones for several reasons. Unlike wide-area cellular phones, short-distance radio LANs operate at low transmit power, typically at less than 100 mW (20 dBm). In addition, transmitting antennas are usually located at more than an arm's length away and not held close to the brain. Hence, much of the radiation will be attenuated before it reaches the user. Finally, cellular phones employ continuous, connection-oriented transmission which is in contrast to the short, intermittent packet transmissions in radio LANs (typically less than 6 ms). While health concerns over radio transmission are common, most people are unaware that popular LAN cables such as UTP-Category 5 contain the chemical halogen which can release toxic fumes in a fire.

2.5   Infrared LANs

The first wireless LAN was developed using infrared transmission some 20 years ago. Since then, numerous infrared LANs have been demonstrated. These systems seek to exploit the many advantages enjoyed by infrared over radio as a medium for wireless transmission. For instance, infrared has an abundance of unregulated bandwidth, is immune to radio interference, and infrared components are small and consume little power.

Infrared LANs use part of the electromagnetic spectrum just below visible light as the transmission medium. Being near in wavelength, infrared light possesses essentially all the physical properties of visible light. Like visible light, infrared light operates at very high frequencies. This implies it travels in straight lines and cannot penetrate opaque objects and physical obstructions (e.g., walls, partitions, ceilings) and will be significantly attenuated passing through windows. This characteristic helps to confine optical infrared energy within a single room, virtually eliminating interference problems and unauthorized eavesdropping. However, infrared light will pass through open doorways, reflect off walls, and bounce around corners just like sunlight and office lighting. Since infrared light has a longer wavelength than visible light, it is invisible to the naked eye under most lighting conditions.

Unlike radio waves, infrared frequencies are too high to be modulated in the same way as radio frequencies. Thus, infrared links are usually based on on-off pulse modulation and detection of the optical signal. On-off pulse transmission is achieved by varying the intensity (amplitude) of the current in an infrared emitter such as a laser diode or a light-emitting diode. In this way, data is carried by the intensity (and not the frequency or phase) of the light wave. Direct detection is performed by a photodiode detector that produces an electrical current proportional to the incident optical power. The detector area determines receiver sensitivity and thus, range. Large-area detectors have high capacitance which limits the available bandwidth. Similar to the many emitters used in optical fiber networks, the infrared emitters employed by infrared LANs operate around the 850 nm wavelengths (nearly 350 THz). However, the high performance of optical fiber networks is largely due to the properties of the fiber. Remove the fiber (as in a wireless system), and the low-loss propagation path is no longer available. Conveying light between two communicating nodes in a controlled, reliable manner then becomes a challenge. Note that infrared systems use two physically different components (e.g., emitters and detectors) to transmit and receive optical signals. This is in contrast to radio systems which typically employ a common antenna for transmission and reception. Thus, as long as the same frequency channel is used, the propagation characteristic between a radio transmitter and a radio receiver is typically symmetrical (reciprocal) but this may not be applicable to optical infrared systems.

2.5.1 Directed and diffuse infrared LANs

Infrared transmission in wireless LANs is either directed (line-of-sight) or diffuse (reflective). In directed infrared LANs, transmitters and receivers have to be aimed at each other to provide line-of-sight communication. The transmitters employ narrowly focused beams while the receivers operate with relatively small viewing angles. Hence, a directed infrared LAN is susceptible to shadowing caused by objects or people positioned between the transmitter and receiver. Most directed infrared LANs provide Ethernet or token ring connectivity. Others employ directional high-powered laser beams for the transmission of high-speed data, voice, and video between buildings. Data rates vary from 1 to 155 Mbps with a range of about 1 to 5 km. High performance directed infrared LANs are primarily used only to implement fixed networks. They are impractical for mobile nodes since the accurate alignment between transmitter and receiver is difficult or impossible for mobile communications.

Diffused infrared LAN systems do not require direct line-of-sight but can only be used indoors since they rely heavily on reflected infrared energy for communication. The infrared signals fill the coverage area like overhead lighting using reflecting surfaces (e.g., walls, partitions, ceilings) to bounce data signals between the transmitter and receiver. This implies a wide field of view which can be achieved by using transmitters consisting of multiple emitters pointed in different directions and by using receivers consisting of arrays of photodiodes. The transmitted infrared signals typically illuminate the ceiling while receivers are pointed at the ceiling to detect infrared energy. In addition, since the signals follow many paths, this provides omnidirectional communication independent of position and orientation of the mobile node's antenna. The advantage of this approach is that a transmitter can communicate with multiple receivers. Shadowing is not a serious problem because light still reaches the receiver due to reflections from the surrounding environment. The disadvantage is that the distance and data rates are reduced as a result of lost infrared energy. Diffusive systems are also more prone to temporal dispersion caused by multipath propagation than directed systems because their larger fields of view imply that more light hits potential reflectors and more reflected light is detected. For the same range, the intersymbol interference caused by multipath is the same, regardless of whether the signal is radio or infrared. However, flat (Rayleigh) fading is generally not a significant impairment for infrared since the extremely short wavelength results in a small spatial extent that affects only a small area of the photodiode detectors [4]. Currently, the highest data rate offered by commercial diffuse infrared LANs is limited to 4 Mbps and these LANs operate within a range of about 10 to 20 m.

2.5.2 Characteristics of infrared LANs

Infrared LANs differ from radio LANs in several ways. In general, radio systems will always afford wider coverage than optical wireless systems because a higher transmitter power can be used and receivers can take full advantage of sensitive heterodyning techniques. On the other hand, radio will always be narrower bandwidth than optical although commercial systems have yet to fully exploit the optical bandwidth.

Power Consumption

Since infrared LANs transmit using on-off pulses, emitters are turned on for a small percentage of the time, thus leading to low overall power consumption. If a greater intensity of radiation is required to extend signal range, the average power can be kept constant by decreasing the duration of the transmitted pulses.

Interference Sources

Infrared communication is immune to radio frequency and electromagnetic interference sources. Conversely, infrared light does not interfere with other communication media. Although common consumer devices such as infrared remote controls also operate in the same optical band as infrared LANs, these devices usually transmit intermittently and hence do not interfere with the proper operation of wireless LANs significantly. Generally, for low to moderate data rates, ambient noise sources that radiate in the same wavelengths as infrared light (e.g., sunlight and artificial light generated by incandescent and fluorescent lamps) are the main factors for the performance degradation of infrared links [5].

Security

Since infrared signals do not penetrate walls and are heavily attenuated by windows, they typically do not propagate to the outside.

Regulation

An infrared wireless LAN can be used anywhere in the world because there are no regulatory restrictions on the operating optical band. Regulatory standards that apply to the use of infrared radiation are safety (as opposed to spectral) regulations.

Health Concerns

The eye is more sensitive to infrared radiation than skin. Thus, high power infrared communications using directed laser beams are governed by safety standards for eye exposure. The emitters used in diffuse infrared LANs radiate less infrared energy than a standard light bulb and hence do not pose any safety hazard.

2.6 Summary

Radio and infrared LANs have their individual strengths and weaknesses. Spread spectrum is by far the most popular radio transmission method in wireless LANs. However, spread spectrum is generally not used as a multiple access technique in wireless LANs. Rather, it is used to protect data signals against the effects of multipath propagation and other impairments. Despite the disadvantages of shadowing and small coverage areas, the abundance of unregulated bandwidth makes infrared an attractive medium for high-speed wireless LANs.

References

[1]   Tuch, B. "Development of WaveLAN, an ISM Band Wireless LAN," *AT&T Technical Journal,* Vol. 72, No. 4, July/August 1993, pp. 27–37.

[2]   Pursley, M. "The Role of Spread Spectrum in Packet Radio Networks," *Proceedings of the IEEE,* Vol. 75, No. 9, January 1987, pp. 116–134.

[3]   David, I. and R. Krishnamoorthy, "Barker Code Position Modulation for High Rate Communication in the ISM Bands," *Bell Labs Technical Journal,* Autumn 1996, pp. 21–40.

[4]   Bantz, D., and F. Bauchot, "Wireless LAN Design Alternatives," *IEEE Network,* March/April 1994, Vol. 8, No.2, pp. 43–53.

[5]   Valadas, R., A. Tavares and A. Duarte, "The Infrared Physical Layer of the IEEE 802.11 Standard for Wireless Local Area Networks," *IEEE Communications Magazine,* Vol. 36, No. 12, December 1998, pp. 107–112.

Wireless LAN Implementation

This chapter discusses practical issues related to wireless LAN implementation. It covers general wireless LAN components and protocol architectures, types of wireless LAN topologies, problems associated with wireless LAN deployment and describes how the performance of an implemented wireless LAN can be enhanced.

3.1 Wireless LAN components

Typical wireless LAN components include wireless network interface cards, wireless access points, and remote wireless bridges.

3.1.1 Wireless network interface cards

Wireless network interface cards are not much different than the adapter cards used for wired LANs. Like wired network adapter cards, the wireless network interface card communicates with the network operating system via a dedicated software driver, thus enabling applications to utilize the wireless network for data transport. Unlike wired adapter cards, however, these adapters do not need a cable to connect them to the network and this allows relocation of network nodes without the need to change network cabling or connections to patch panels or hubs. Typical wireless network interface cards for notebook computers and personal computers are shown in Figure 3.1 and 3.2 respectively.

3.1.2 Wireless access points

Access points create wireless coverage areas that connect mobile nodes to existing wired LAN infrastructures. This enables a wireless LAN to become an extension of a wired network. Because access points enable extension of a wireless coverage zone, wireless LANs are inherently very scalable and additional access points can be deployed throughout a building or campus to create large wireless access zones. The access points not only provide communication with the wired network but can also filter traffic and perform standard bridging functions. The filtering function helps to conserve bandwidth on the wireless link by removing redundant traffic (see Section 3.5.5). Due to the bandwidth mismatch between the wireless and wired media, it is important for an access point

Figure 3.1 Wireless network interface card for notebook computers (photo courtesy of Lucent Technologies).

Figure 3.2 Wireless network interface card for personal computers (photo courtesy of Lucent Technologies).

to have adequate buffer and memory resources. Buffers are also essential for storing data packets at the access point when a mobile node temporarily moves out of a wireless coverage area or when a mobile node operates in the low-power (sleep) mode (see Section 3.5.10). Access points communicate with each other through the wired network in order to manage the mobile nodes. An access point need not control access from multiple mobile nodes, (i.e., it can operate with a distributed random access protocol such as CSMA). However, a centralized multiple access protocol controlled by an access point does offer some advantages (see Sections 3.3 and 3.4.1). Common wired network interface options to an access point include 10Base2, 10BaseT, cable modem, ADSL, ISDN and modem. A typical wireless access point is shown in Figure 3.3. Some wireless network interface cards may be used in conjunction with wireless access points.

3.1.3   Remote wireless bridges

Remote wireless bridges are similar to access points except that they are primarily used for outdoor links. Depending on the distance and coverage, external antennas may be required. Such bridges are designed to link networks together, typically in different buildings and as far as 32 km (12 miles) apart. They offer a quick and low-cost alternative to installing cable or leased telephone lines and are often used when traditional wired interconnections are impractical (e.g., rivers, rough terrain, private property, and highways). Unlike cable links and dedicated telephone circuits,

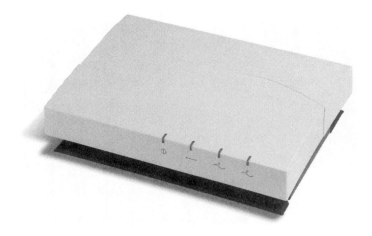

Figure 3.3 Wireless access point (photo courtesy of Lucent Technologies).

wireless bridges are able to filter traffic and ensure that the connected networks are not overwhelmed with unnecessary traffic. These bridges may also serve as internal security devices since they read only addresses encoded into LAN adapters (i.e., the MAC addresses), thus making successful spoofing (fraudulent communications) attempts exceedingly difficult. A typical remote wireless bridge is shown in Figure 3.4.

3.2 Wireless LAN protocol architectures

Wireless LANs differ from conventional wired networks primarily at the physical layer and at the medium access control (MAC) sublayer of the open systems interconnection (OSI) reference model. These differences give rise to two approaches in providing the logical interface point for wireless LANs. If the logical interface point is at the logical link control (LLC) layer, this approach usually requires custom drivers to support higher level software such as the network operating system [1]. Such an interface allows mobile nodes to communicate directly with each other using wireless network interface cards. The other logical interface point is at the MAC sublayer and is typically employed by wireless access points.

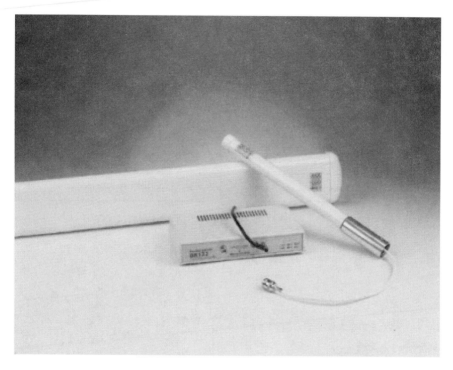

Figure 3.4 Remote wireless bridge (photo courtesy of Lucent Technologies).

For this reason, wireless access points perform bridging and not routing functions. Although a MAC interface requires a wired connection, it permits any network operating system or driver to work with the wireless LAN. Such an interface enables an existing wired LAN to be extended easily by providing access for new wireless network devices. The protocol architectures of typical wireless LAN network interfaces are shown in Figure 3.5. The lower layers of a wireless interface card are usually implemented on firmware that runs on embedded processors. Higher layers of the network protocol stack are provided by the operating system and application programs. A network driver allows the operating system to communicate with the lower layer firmware embedded in the wireless network interface card. In addition, it performs standard LLC functions. For the Windows operating system, the driver generally complies with some version of the network driver interface specification (NDIS). Drivers based on Unix, Linux, and Apple Powerbook are also available.

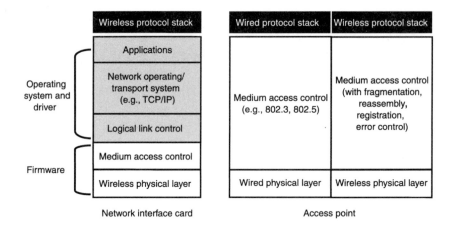

Figure 3.5 Protocol architectures of wireless LAN components.

3.3 Wireless LAN topologies

Wireless LANs often adopt two types of network configurations or topologies. These are the independent and infrastructure topologies as depicted
in Figures 3.6 and 3.7 respectively. An independent topology supports
peer-to-peer connectivity where mobile nodes communicate directly
with each other using wireless adapters. Because these ad-hoc networks
can be implemented quickly and easily, they are usually created without
the need for special tools or skills. They also require no network administration. Such configurations are ideal in business meetings or in the
setting up of temporary workgroups. However, they may suffer the disadvantage of limited coverage area. An access point can extend the range
of two independent wireless LANs by acting as a repeater, effectively
doubling the distance between mobile nodes.

Infrastructure wireless LANs allow mobile nodes to be integrated into
the wired network (see Figure 3.7). The transition from the wireless to
the wired media is via an access point. The design of a wireless LAN can
be simplified considerably if information about the network and the
intelligence to manage it are all collected in one location. A centrally
located access point can control and arbitrate access among contending
nodes, provide convenient access to the backbone network, assign ad-

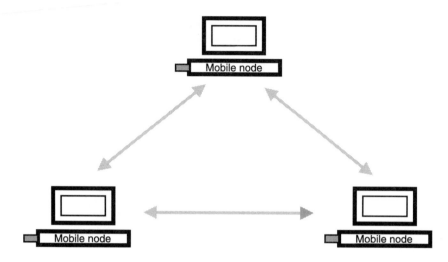

Figure 3.6   Independent (ad-hoc) wireless LAN.

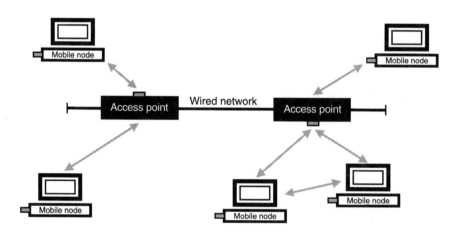

Figure 3.7   Infrastructure wireless LAN.

dresses and priority levels, monitor network load, manage forwarding of packets, and keep track of the current network topology [4]. However, a centralized multiple access protocol does not allow a node to transmit directly to another node that is located within the coverage area of the same access point (see Figure 3.8). In this case, a packet will have to be transmitted two times (first from the originating node and then from the

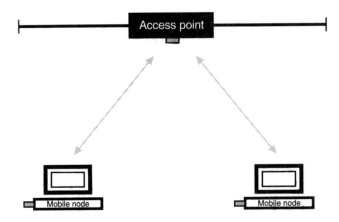

Figure 3.8 Infrastructure wireless LAN under centralized control.

access point) before it reaches the destination node, a process that reduces transmission efficiency and increases propagation delay. Nevertheless, such systems generally provide higher data throughputs, larger coverage areas, and are capable of servicing time-sensitive traffic involving voice and video. In addition, a strategically-located access point may also minimize transmit power and deal with the problem of hidden nodes (see Section 3.4.1) effectively. Note that since most wireless LANs employ distributed protocols such as CSMA for multiple access, it is possible for nodes in an infrastructure network to communicate directly with each other (see Figure 3.9). However, some infrastructure wireless LANs require packet transmission to be addressed only to the access point even when CSMA is adopted. The access point will then relay the packets to the correct destination node.

3.4 Wireless LAN deployment considerations

In this section, the major issues related to the deployment of wireless LANs are discussed. These are the hidden node problem, power capture, radio interference sources, and obstructions to signal propagation. Most of these issues pertain more to radio LANs.

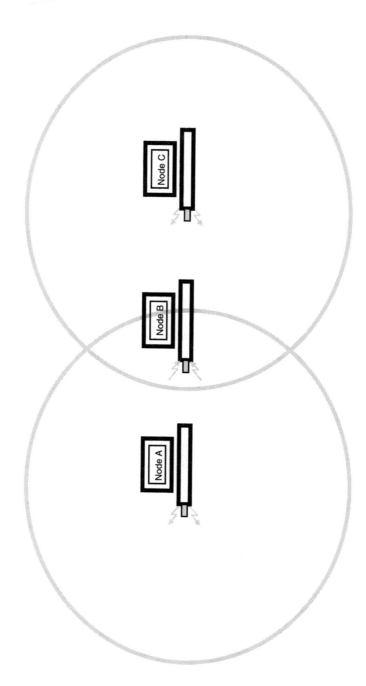

Figure 3.9 Hidden node problem for an independent wireless LAN.

3.4.1 Hidden node

A difficulty with large signal power fluctuations in wireless LANs is the existence of hidden (nonsited) nodes where some nodes are within the range of the intended receiver but not of the transmitter. This makes coordination among distributed nodes an arduous task since the wireless link is no longer capable of pure broadcast, and transmission from one node may be detected by an arbitrary number of other nodes. As an example, Figure 3.9 shows two nodes (A and C) that are within receiving range of a third node B. However, nodes A and C are not in range. If nodes A and C attempt to transmit to node B simultaneously, node B will experience a collision and will not be able to receive either transmission. Both A and C will not be aware of the collision. Carrier sensing is rendered ineffective in a hidden-node situation since a source node inhibits the other nodes within its vicinity rather than those in the vicinity of the destination node. This in turn degrades the performance of carrier sensing protocols since the vulnerable collision duration now covers the entire length of the data packet. With normal carrier sensing, the vulnerable period is much shorter, typically within the first few bits of a data packet.

Hidden nodes are not a problem if wireless coverage areas are well isolated. Because collisions are less common in spread spectrum systems than narrowband systems (see Section 2.4.3), the existence of hidden nodes may not pose a serious problem for DSSS and FHSS wireless LANs. On the contrary, hidden nodes may benefit both systems since without carrier sensing, multiple packet transmission using different time-shifted versions of a common pseudonoise code or hopping pattern may now coexist (see Figure 3.10).

Figure 3.11 shows how hidden node collisions can also occur in an infrastructure wireless LAN. In this case, the access point experiences a collision due to overlapping transmission from both nodes A and B. A more severe problem here is that node B may not be able to communicate with node A if the access point is not configured as a repeater that relays packet transmission among nodes within its coverage area. A centralized multiple access protocol coordinated by the access point solves the hidden node problem for an infrastructure LAN. Nodes cannot transmit unless explicit permission is given by the access point. However, such a protocol collision may still occur when two adjacent access points transmit simultaneously to a node in an overlapped region (Figure 3.12). This

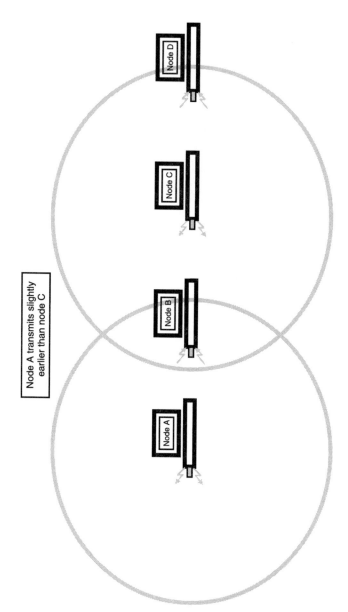

Figure 3.10 Multiple transmissions in a hidden node situation.

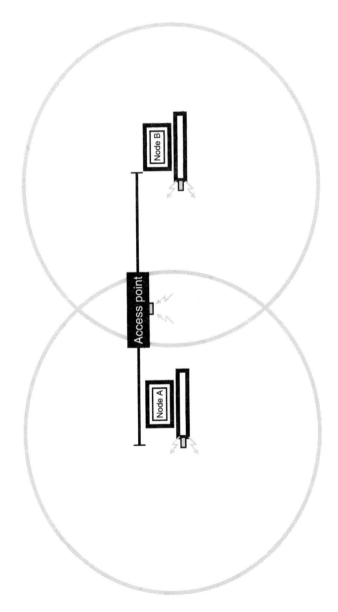

Figure 3.11   Hidden node problem in an infrastructure wireless LAN.

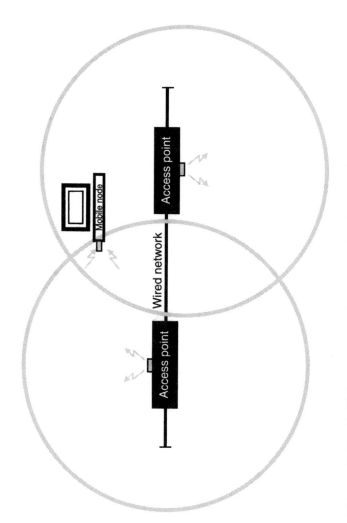

Figure 3.12   Collision in an overlapped region for an infrastructure wireless LAN under centralized control.

situation can be alleviated if adjacent access points coordinate transmission through the wired network or operate using nonoverlapping
frequency channels.

3.4.2 Power capture

The large changes in signal attenuation admit the possibility of power
capture that allows a radio receiver to successfully decode a stronger
signal even when more than one node happens to transmit at the same
time. This is because radio receivers are capable of tracking the strongest
of many signals if the power of the next-to-strongest is down by 1.5 to
3 dB [2]. Distance is a major factor that determines the signal power
received.

Suppose two nodes A and C are attempting to communicate with
node B (Figure 3.13). Both nodes are within range of node B. However,
since node A is closer to node B, the radio signal received from node A
can overpower the signal from node C if the transmission from both nodes
overlap in time. This raises a fairness issue since the node furthest away
is always discriminated and there is a possibility that node C will never
be able to communicate with node B. On the other hand, the capture
effect can help to reduce the probability of collisions (including hidden
node collisions) and hence improve the network performance of the
wireless LANs.

In spread spectrum systems, capture refers to the ability of a receiver
to successfully decode a packet with a given pseudonoise code or hopping
pattern despite the simultaneous presence of other time-overlapping
signals with the same code or hopping pattern. Usually, the packet

Figure 3.13   Power capture.

captured by the receiver corresponds to either the first arriving signal (time capture) or the strongest signal (power capture). In general, power capture does not occur for FHSS systems unless two or more transmitting nodes employ a common hopping pattern and the frequency channels are exactly synchronized in time. However, most FHSS wireless LANs operate with common hopping patterns and synchronized frequency channels. Power capture can pose a serious problem for DSSS systems since a packet with a stronger signal can overwhelm the interference rejection capability (i.e., the processing gain) of a DSSS system. For DSSS systems involving CDMA, power control becomes even more critical since multiple user transmissions typically overlap time. The IEEE 802.11 standard mandates the use of power control for both DSSS and FHSS transmission with power levels above 100 mW. Although such control allows power to be utilized efficiently, it can be difficult to maintain in a fading or highly mobile environment.

3.4.3 Radio interference sources

For wireless LANs that operate in the 2.4 GHz radio frequency band, microwave ovens can be an important source of interference. The ovens emit up to 750 W of power at 50 pulses per second with a radiation cycle of about 10 ms. Hence, for a wireless data rate of 2 Mbps, the maximum packet length is restricted to about 20,000 bits or 2,500 octets. The emitted radiation sweeps from 2.4 to 2.45 GHz and remains stable for a short period at 2.45 GHz. Even though the units are shielded, a good amount of energy can still interfere with the transmission from wireless LANs.

Other sources of interference in the 2.4 GHz frequency band include photocopiers, theft-detection devices, elevator motors, and medical equipment.

3.4.4 Obstructions to signal propagation

As explained in Section 2.5, infrared signals are blocked by opaque objects and physical obstacles, and are significantly attenuated by glass windows. For radio signals, how far these signals will travel depends largely on the construction materials of the walls, partitions, and other objects (see Table 3.1).

Table 3.1
Radio Propagation Barriers and Their Effects

Barrier	Level of Attenuation	Examples
Plasterboard	Low	Inner Walls
Wood/Synthetic Material	Low	Partition
Asbestos	Low	Ceilings
Glass	Low	Windows
Water	Medium	Aquarium
Bricks/Marble	Medium	Inner/Outer Walls
Concrete	High	Floors
Metal	Very High	Steel Cabinets/ Reinforced Concrete

3.5 Enhancing the performance of wireless LANs

This section examines several methods that enhance the performance of wireless LANs. Specifically, techniques such as increasing network capacity with multiple frequency channels, extending wireless coverage area with data rate fallback, filtering redundant traffic, providing mobility support through roaming, improving network congestion with load balancing, and securing network access shall be discussed.

3.5.1 Multichannel configuration

Multichannel configurations may prove very useful in environments with a high concentration of wireless nodes operating in the same vicinity. If a particular coverage area of a wireless LAN has more nodes and requires additional bandwidth, a second access point that operates a different frequency channel can be added, thereby doubling available bandwidth. Multichannel operation also allows access points to service nodes with high-speed demands and is only applicable to radio LANs. By configuring different access points with different frequency channels, transmission in one wireless coverage area becomes isolated from another, thus reducing mutual interference and the frequency of nodes deferring communications. For a single channel system, nodes in the shaded region shown in Figure 3.14 share the same medium. This means if one of the nodes in the region transmits, all other nodes must defer communication. By assigning each access point with a different channel, congestion in the region is reduced due to the spreading of traffic load between two access points. Independent networks do not support multi-channel operation.

Multichannel operation can also be applied to wireless bridging (see Figure 3.15). Since a different frequency channel is used for bridging, it will not interfere with normal access point operation. This enables the range of a wireless LAN to be extended without the need for a wired backbone. Some wireless LANs require a separate access point for wireless bridging while others require directional outdoor antennas.

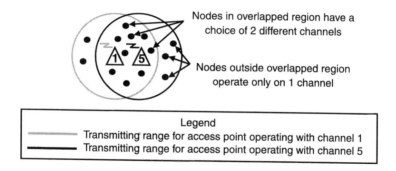

Figure 3.14   Multichannel operation.

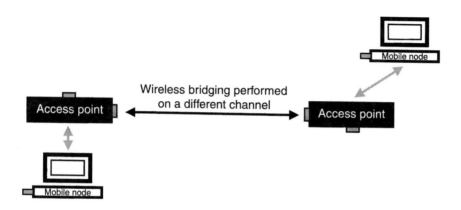

Figure 3.15   Multichannel wireless bridging for range extension.

3.5.2 Multichannel operation for 2.4 GHz DSSS wireless LANs

In the 2.4 GHz ISM band, the total allocated bandwidth for DSSS wireless LANs can be separated by different carrier frequencies. The number of selectable carrier frequencies is as follows: North America 11; most of Europe 13; France 4; Japan 1. Since the DSSS signal is spread over a wide bandwidth, the preferred carrier frequency separation between neighboring access points is at least 30 MHz (see Chapter 2, Figure 2.10). This means that in the United States and most of Europe, up to three carrier frequencies can be applied within the same region. To illustrate this, Table 3.2 shows 13 possible DSSS multichannel assignments based on 13 different carrier frequencies. Allowing the maximum carrier frequency separation will decrease the amount of adjacent interference and provide a noticeable performance enhancement over networks with minimal separation.

Table 3.2
Multichannel Assignment for 2.4 GHz DSSS Wireless LANs [3]

Assignment	Access Point 1	Access Point 2	Access Point 3
1	2412 MHz (1)	2472 MHz (13)	2442 MHz (7)
2	2417 MHz (2)	2472 MHz (13)	2442 MHz (7)
3	2422 MHz (3)	2472 MHz (13)	2447 MHz (8)
4	2427 MHz (4)	2472 MHz (13)	Not Applicable
5	2432 MHz (5)	2472 MHz (13)	Not Applicable
6	2437 MHz (6)	2472 MHz (13)	2412 MHz (1)
7	2442 MHz (7)	2412 MHz (1)	2472 MHz (13)
8	2447 MHz (8)	2412 MHz (1)	2472 MHz (13)
9	2452 MHz (9)	2412 MHz (1)	Not Applicable
10	2457 MHz (10)	2412 MHz (1)	Not Applicable
11	2462 MHz (11)	2412 MHz (1)	2437 MHz (6)
12	2467 MHz (12)	2412 MHz (1)	2442 MHz (7)
13	2472 MHz (13)	2412 MHz (1)	2442 MHz (7)

Note: Carrier frequency numbers are indicated in parentheses.

3.5.3 Multichannel operation for 2.4 GHz FHSS wireless LANs

Since the frequency channels in a hopping pattern occupy the entire 2.4 GHz ISM band, the channelization method used in DSSS wireless LANs cannot be directly applied to FHSS systems. FHSS wireless LANs accomplish multichannel operation by implementing separate channels on different hopping patterns.

3.5.4 Fall back rate

Most wireless LANs take advantage of short-range and good propagation conditions to increase the data rate. While signal transmission at a lower data rate is usually more reliable and permits a greater coverage area, a higher throughput performance may sometimes be preferred. To balance speed against coverage area, the wireless network interface card normally transmits at the maximum available data rate. When a data transmission fails more than once, the interface card will fall back to a lower rate.

3.5.5 Filtering network traffic

One of the ways to optimize the performance of a wireless LAN is to prevent redundant traffic from being transmitted over the wireless link. Redundant traffic may include:

- Network messages exchanged by wired networking devices (e.g., servers) that are not relevant to the wireless clients;

- Broadcast/multicast messages that are not specifically addressed to the wireless clients;

- Error messages that are generated by malfunctioning devices or by incorrectly configured devices (e.g., devices in closed network loops).

Filtering redundant traffic will save the bandwidth of the wireless link for the mobile nodes. This can be achieved using an access point's bridging functions:

▶ Protocol filtering to deny wired networking protocols from being bridged to the wireless interface;

▶ Filtering traffic exchanged between two specific nodes;

▶ Enabling the spanning tree mechanism to resolve closed network loops errors;

▶ Storm threshold filtering to limit the number of messages being bridged.

3.5.6 Roaming and handoff

A key requirement for wireless LANs is the ability to handle mobile and portable nodes. A portable node is one that is moved from location to location but is only used while in a fixed location. Mobile nodes actually access the LAN while in motion. User mobility requires a roaming function that enables a mobile node to migrate between different physical locations within the LAN environment without losing the network connection. To allow seamless roaming, each of these locations must typically be serviced by an access point and the wireless coverage areas of the access points must be properly overlapped. A mobile node will monitor the signal-to-noise ratio (SNR) as it moves and, if required, scan for available access points and then automatically connect to the desired access point to maintain continuous network access (see Figure 3.16). Usually, the

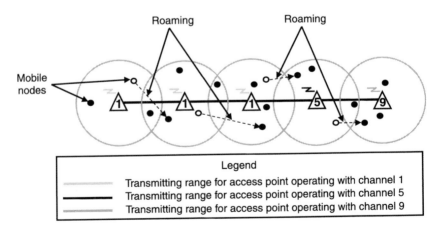

Figure 3.16   Roaming in wireless LANs.

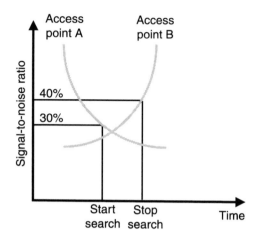

Figure 3.17   Searching for access points during roaming.

SNR is a function of both signal strength and signal quality. When the SNR falls below a predetermined threshold, the node will search for a nearby access point with the better SNR (see Figure 3.17). If such an access point is detected, the mobile node sends a handoff request to the new access point which will in turn relay the request to the old access point (see Figure 3.18). The old access point will release control of the ongoing connection and transfer it to the new access point. The handoff is completed when the mobile node is notified of the change. This procedure is similar to the handoff function in cellular networks, with one major difference—roaming on a packet-based wireless LAN is easier since the transition from one coverage area to another may be performed between packet transmission, as opposed to cellular telephony where the transition may occur during a voice connection. It is crucial that handoffs be executed quickly since the high data rates of wireless LANs imply that many packets can be in transit while a handoff is being negotiated. This may result in excessive retransmission due to lost or misdirected packets. The rate at which handoffs are triggered is highly dependent on the rate at which the SNR is deteriorating.

Most wireless LANs are able to support mobile nodes with pedestrian speeds (e.g., up to 10 km/h). Some wireless LANs are able to guarantee uninterrupted network connectivity without losing or duplicating frames even when nodes move from one coverage area to another at speeds of

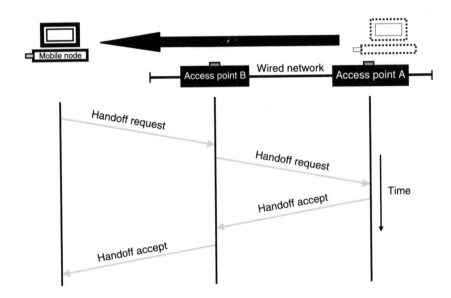

Figure 3.18   Handoff negotiation.

60 km/h. To support roaming in multichannel configurations, mobile nodes are typically capable of switching frequency channels or hopping patterns automatically when roaming between access points. Independent networks usually do not support roaming.

3.5.7   Mobile Internet support

The World Wide Web (WWW) is perhaps the largest source of distributed information. The network that drives the Web is the Internet. While the size of the Internet has been increasing at a rapid rate, the use of the Web has grown at an explosive pace. With the advent of dynamic content (e.g., HTML, CGI, XML) and executable content (e.g., Java), the Web is poised to play a central role in the process of making information ubiquitously accessible. The Internet runs on internetworking devices called routers that forward network layer Internet Protocol (IP) packets among networks with different link layers. Managing the network layer is much more flexible than the link layer since network administrators can assign structured IP addresses as opposed to unstructured MAC (link layer) addresses. However, mobile computing poses a problem if nodes with fixed IP addresses are moved from one location to another. Although a wireless LAN provides mobility support with roaming services, true mobility for large IP networks can only be realized if the IP addresses are assigned dynamically. This is because IP addresses are location dependent and therefore need to be changed to reflect the different locations of a mobile node. Hence, the problem of mobility in IP networks lies with the way IP addresses are structured, with or without wireless connectivity. Dynamic allocation of IP addresses is usually applied to mobile nodes and not to wireless access points since the access points occupy static (fixed) locations.

The dynamic host configuration protocol (DHCP) is one method of configuring IP addresses dynamically regardless of location. A dynamically assigned IP address is known as an active lease. The active lease usually has an expiration date, which allows automatic reallocation of IP addresses that are no longer in use. Thus, DHCP relieves the administrative burden of managing IP addresses in addition to providing mobile IP addresses. When DHCP is implemented on a wired IP network, a mobile client is able to connect to the network in different locations (e.g., in different subnets). This is achieved by physically disconnecting the network cable from a fixed outlet or socket and reconnecting at the new site. Hence, ongoing connections will have to be broken when a client on a wired network moves to a new location. However, when DHCP is applied to wireless LANs, IP connections can be maintained (and applications can continue to run) even as a mobile client changes location. This removes the need to log in and out of the network. Thus, wireless LANs with mobile IP addresses can provide continuous and location independent access to Internet services.

3.5.8 Load balancing

Load balancing allows wireless LANs to serve greater loads more efficiently. Each access point can monitor the traffic load within its coverage area and then try to balance the number of nodes serviced according to the traffic load in adjacent access points. To achieve this, access points must exchange traffic load information through the backbone network. Most load balancing methods do not depend on signal strength as this may complicate the roaming algorithm. Typically, roaming has priority over load balancing since a mobile node must first be able to connect to an access point with reasonable signal quality/strength before load balancing can be performed.

3.5.9 Securing wireless access

The free-space wireless link is more susceptible to eavesdropping, fraud, and unauthorized transmission than its wired counterpart. Being an open medium with no precise bounds makes it impractical to apply physical security like in wired networks. Nevertheless, several alternative security mechanisms can be used to prevent unauthorized access of data transmitted over a wireless LAN. These are:

▶ Encrypting all data transmitted via the wireless link;

▶ Closing the network to all nodes that have not been programmed with the correct network identification;

▶ Restricting access within a wireless LAN by listing only those nodes that are allowed to transmit data;

▶ Implementing passwords in network operating systems.

3.5.10 Power management

Portable wireless LAN devices rely on limited battery power to conduct communications over a wireless link that is prone to error bursts due to fading and other propagation impairments. As current battery research does not predict a substantial change in the available energy in a consumer battery, it is crucial that wireless mobile devices are designed to be efficient in energy usage. Minimizing energy usage is a significant constraint since it impacts design at all levels of network control. There has been substantial research in energy-efficient hardware (e.g., low power electronics, processor sleep-time, and power-efficient modulation) for mobile communications. However, due to fundamental physical limitations, progress towards further efficiency will become mostly a software-level issue. In wireless LANs, the software protocol can be designed to allow an idle network device to turn off its receiver most of the time, thereby saving considerable amounts of power without compromising performance. More details can be found in Chapter 4.

3.6 Wireless LAN applications and benefits

The most critical factors which will determine the success of wireless LANs are the utility and convenience of deploying end-user devices. Generally, users have higher expectations for in-building than outdoor wireless networks. Wireless LANs are widely applied in the manufacturing, warehousing, retail, health care, transportation, and financial industries (see Table 3.3). In these environments, where workers are mobile for a large percentage of the time and the need for real-time access to constantly changing information is vital, wireless LAN technology can lower costs and increase productivity significantly. An even more important advantage is that wireless LANs enable portable networks, thereby allowing LANs to move with the knowledge that workers need them. In addition, wireless LANs can complement wired LANs in office environments by supporting temporary workgroups, reducing the need for network rewiring and enabling user mobility throughout a building without loss of network connectivity. On the consumer front, the increased number of

Table 3.3
Wireless LAN Applications

Industry	Applications
Retail	Portable point-of-sale, wireless order entry
Financial	Replicated branches, temporary audit workgroups
Medical	Mobile nursing stations, patient record tracking
Transportation	Remote mobile customer service
Education	Mobile classrooms
Manufacturing	Real-time data collection, inventory management
Government	Wireless office automation
Residential	Personal area networks, wireless home networks
Warehousing	Networking forklift trucks

multiple-PC homes, coupled with the growing popularity of residential cable networks, have created a new demand for wireless LANs.

The main benefits of wireless LANs can be summarized as follows:

- Flexible network deployment and extension;
- Ease of installation in hard-to-wire buildings (e.g., hospitals, factories, museums, libraries);
- Reduced installation time;
- Accurate location information of a mobile node;
- Inherent support for broadcast and multicast services;
- Bridging between colocated buildings;
- Wireless and mobile access to existing wired LANs;
- Continuous access to networked resources, including the Internet;
- Long-term cost savings.

3.7 Summary

Wireless LANs find applications in almost any environment—industrial, government, and residential. A basic concern of radio transmission is that unauthorized people can tap the signal from outside. Hence, user access to a wireless LAN must be properly secured. Radio emissions can also be an unintentional source of interference to other wireless networks and must be controlled. Wireless LAN topologies range from small independent networks, suitable for temporary configurations, to infrastructure networks that offer fully distributed connectivity with roaming. Various techniques such as traffic load balancing, power management, and multichannel operation can be employed to enhance the performance of wireless LANs.

References

[1]   Mathias, C., "New LAN Gear Snaps Unseen Desktop Chains," *Data Communications International*, March 1994, pp. 75–80.

[2]   Roberts, L., "Aloha Packet System, With and Without Slots and Capture," *Computer Communications Review*, Vol. 5, No. 2, April 1975, pp. 28–42.

[3]   Lucent Technologies, *User's Guide to WavePOINT-II*, November 1998.

[4]   Bing, B. and R. Subramanian, "A New Multiaccess Technique for Multimedia Wireless LANs," *Proceedings of the IEEE GLOBECOM*, Phoenix, Arizona, November 1997, pp. 1318–1322.

Contents

Wireless LAN Standards

In 1990, the Institute of Electrical and Electronics Engineers (IEEE) formed a committee to develop a standard for wireless LANs operating at 1 and 2 Mbps. In 1992, the European Telecommunications Standards Institute (ETSI) chartered a committee to establish a standard for high performance radio LANs (HIPERLAN) operating in the 20 Mbps range. Recently, wireless LAN standards targeted for specialized applications in the home have emerged. Unlike these standards, the development of the IEEE 802.11 standard was heavily influenced by existing wireless LAN products already available in the market. Hence, although the standard took a relatively long time (nearly 7 years) to complete (due to numerous competing proposals from different vendors), it is by far the most popular standard to date. This chapter explores the various wireless LAN standards with emphasis on the IEEE 802.11 standard.

4.1 The IEEE 802.11 wireless LAN standard

The IEEE 802.11 wireless LAN standard is sponsored by the 802 Local and Metropolitan Area Networks Standards Committee (LMSC) of the IEEE Computer Society. The standard evolved from 6 draft versions and the final draft was approved on June 26, 1997. The standard allows multiple vendors to develop interoperable LAN products for the globally available 2.4 GHz industrial, scientific, and medical (ISM) band. Work is progressing for the standard to achieve recognition as a joint ISO/IEC and IEEE standard.

The IEEE 802.11 standard specifies wireless connectivity for fixed, portable, and moving nodes in a geographically limited area. Specifically, it defines an interface between a wireless client and an access point, as well as among wireless clients. As in any IEEE 802.x standard such as 802.3 (CSMA/CD) and 802.5 (token ring), the 802.11 standard defines both the physical (PHY) and medium access control (MAC) layers. However, the 802.11 MAC layer also performs functions that are usually associated with higher layer protocols (e.g., fragmentation, error recovery, mobility management, and power conservation). These additional functions allow the 802.11 MAC layer to conceal the unique characteristics of the wireless PHY layer from higher layers.

4.1.1 IEEE 802.11 network architecture and reference model

The basic service set (BSS) is the basic building block of an 802.11 wireless LAN and consists of two or more mobile nodes (called stations or STAs). Figures 4.1 and 4.2 illustrate the concept of the BSS when applied to independent and infrastructure wireless LANs. Each BSS has an identification called the BSSID that normally corresponds to the MAC address of the wireless portion of the network interface card. The wireless coverage area within which members of a BSS can communicate is known as the Basic Service Area (BSA). An independent wireless LAN comprises only one BSS and is known as an independent BSS (IBSS). A distribution system (DS) interconnects two or more BSSs together, usu-

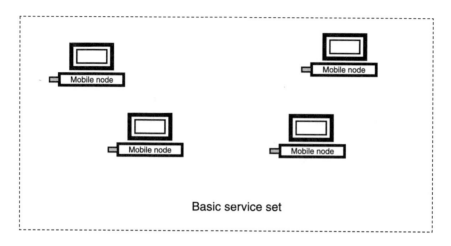

Figure 4.1 Basic service sets in an independent network.

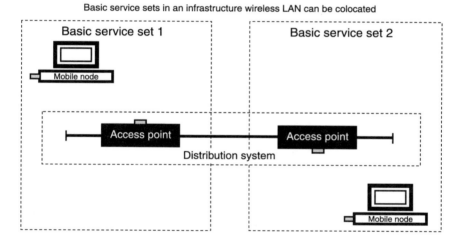

Figure 4.2 Basic service sets in an infrastructure network.

ally through a wired backbone network, thereby allowing mobile nodes to access fixed network resources. A wireless LAN comprising, a collection of BSSs and the DS is known as an extended service set (ESS). Like the BSS, an ESS also has a unique identification called the ESSID. Defining a common ESSID allows a mobile node to roam from one BSS to another.

4.1.2 The IEEE 802.11 basic reference model

As shown in Figure 4.3, the PHY layer is divided into two sublayers. The physical medium dependent (PMD) sublayer deals with the characteristics of the wireless medium (i.e., DSSS, FHSS, or DFIR) and defines a method for transmitting and receiving data through the medium (e.g., modulation and coding). The physical layer convergence procedure (PLCP) sublayer specifies a method of mapping the MAC sublayer protocol data units (MPDUs) into a packet format suitable for the PMD sublayer. It can also perform carrier sensing (channel assessment) for the MAC sublayer. The MAC sublayer defines a basic access mechanism (based on CSMA) for multiple mobile nodes to access the wireless medium (see Section 4.4). It can also perform fragmentation and encryption of data packets. The PHY layer management is concerned with adapting to different link conditions and maintaining a PHY layer management information base (MIB). The MAC sublayer management deals with synchronization, power management, association, and reassociation issues. In addition, it maintains the MAC sublayer MIB. Finally, station management specifies how the PHY and MAC management layers interact with each other.

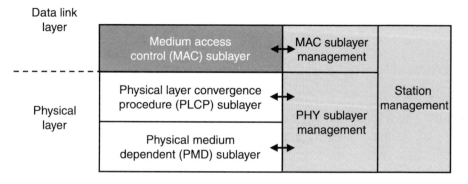

Figure 4.3 IEEE 802.11 basic reference model.

4.2 IEEE 802.11 physical layer

The PHY layer specification allows three transmission options that enable 802.11 wireless LANs to be deployed in different coverage areas ranging from a single room to an entire campus. These options are direct-sequence spread spectrum (DSSS), frequency-hopping spread spectrum (FHSS) and diffuse infrared (DFIR). However, in order for 802.11 wireless devices to interoperate, they have to conform to the same PHY layer (i.e., FHSS wireless LANs will communicate with other FHSS wireless LANs, but not with DSSS wireless LANs). While the DFIR PHY layer operates at base-band, the two radio frequency options (i.e., DSSS and FHSS) operate at the 2.4 GHz ISM band. This frequency band does not require the user to have a specific license although the vendor needs to obtain a type license for the country where it sells the products. The DSSS 802.11 specification supports mandatory data rates of 1 and 2 Mbps. For the FHSS and DFIR specifications, the 1 Mbps data rate is mandatory while the 2 Mbps data rate is optional. Each PHY layer specification is formally described using state diagrams.

4.2.1 General packet formats

User information is segmented into data packets (802.11 prefers the term "frames") with a PLCP preamble and a PLCP header attached to the start of each packet. After a receiving node synchronizes to the PLCP preamble, it obtains the length of the data packet, the data rate (e.g., 1 or 2 Mbps), and other information from the PLCP header. It is important to note that the PLCP preambles and headers are transmitted at a data rate of 1 Mbps (with the exception of some portions of the DFIR PLCP header). This allows a lower-rate (but longer range) 802.11 wireless LAN to inter-operate with a higher-rate (but shorter range) counterpart. At the same time, the relatively low rate of 1 Mbps enables the PLCP preambles and headers to be decoded without the use of power-hungry equalizers. Such equalizers are usually required to combat the multipath problem at high data rates (see Section 1.6.5). A disadvantage of the 1 Mbps data rate is that it reduces the transmission efficiency when the MPDU is transmitted at a higher data rate (see Section 5.2.1).

4.2.2 DSSS physical layer

The DSSS 802.11 packet format is shown in Figure 4.4. Some of the terms in the various fields of the PLCP header have been expanded to clarify their use. Besides allowing a receiving node to detect the auto-correlation peaks of the pseudonoise code and lock on to the timing of an incoming packet, the synchronization bits also enable selection of the appropriate antenna (if antenna diversity is employed). The signal field indicates whether the MPDU is modulated using DBPSK (1 Mbps) and DQPSK (2 Mbps) and can be used to identify higher data rate extensions. The start frame delimiter indicates the start of the data packet. The length field defines the length of the MPDU while the header error check protects the three fields in the PLCP header.

The 1 Mbps basic data rate uses differential binary phase shift keying (DBPSK) where each data bit is mapped into 1 of 2 phases. The 2 Mbps enhanced rate uses differential quadrature phase shift keying (DQPSK). In this case, 2 data bits are mapped into 1 of 4 phases of the spreading code. Table 4.1 shows the phase definitions for DBPSK and DQPSK. Figure 4.5 illustrates the operation of these modulation schemes. With differential phase shift keying, the information is encoded based on the phase difference between adjacent data symbols. In other words, the transmitted phase (ϕ_n) of the symbol is a function of the previous phase (ϕ_{n-1}) and the phase change $(\Delta\phi)$ as follows: $\phi_n = \Delta\phi + \phi_{n-1}$. Differential phase reception minimizes acquisition time. The DSSS 802.11 specification requires both data rates to be implemented. The receiver input signal level

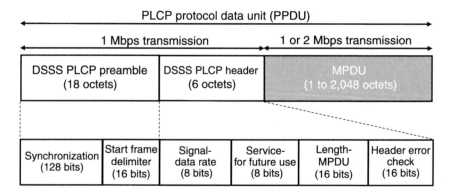

Figure 4.4 DSSS PLCP packet format.

Table 4.1
Phase Definitions of DBPSK and DQPSK

Modulation	Data	Phase Change
DBPSK	0	0°
	1	180°
DQPSK	00	0°
	01*	90°
	11	180°
	10*	270°

* Leftmost bit is transmitted first.

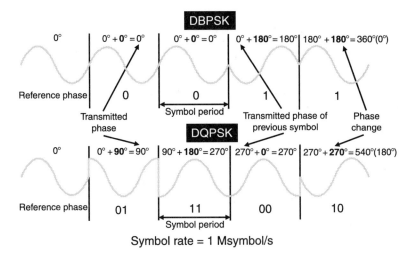

Figure 4.5 DBPSK and DQPSK modulations.

is specified as –80 dBm for a packet error rate of 8×10^{-2}. The packet error rate is the probability of not decoding all the bits in a data packet correctly. It is obtained from the product of the bit error rate and the packet length.

An 11-chip Barker code has been chosen as the pseudonoise code for several reasons. First, it exhibits good autocorrelation properties. Second, because this Barker code is relatively short in length, it allows fast synchronization. Third, the sidelobes are unity bounded, independent of the input information polarity and delay and the low sidelobes imply that less signal power is lost when only the main lobe is accepted (see Figure 2.6, Chapter 2). For each data symbol transmitted, the 11-chip Barker code changes phase 6 times (see Figure 2.2, Chapter 2). It is asymmetric because the number of positive and negative pulses differ by one (a symmetric code contains an equal number of positive and negative pulses). Hence, the MPDU is scrambled to limit dc offset changes due to the asymmetric Barker code. The chip rate of 11 Mchip/s gives rise to a chip period of 90.9 ns. This means that multipath propagation can still pose a problem if the rms delay spread is less than 90.9 ns. Hence, antenna diversity may still have to be employed to combat the undesirable effects of multipath. A general rule of thumb for DSSS systems is that the bandwidth is at least two times the chip rate. Therefore, a chip rate of 11 Mchip/s requires a minimum bandwidth of 22 MHz.

4.2.3 FHSS physical layer

The FHSS 802.11 packet format is shown in Figure 4.6. By comparing the DSSS and FHSS PLCP packet formats, it can be observed that FHSS requires less bits for synchronization. However, the maximum length of the MPDU for FHSS is shorter compared to DSSS.

The 1 Mbps basic data rate uses 2-level Gaussian frequency shift keying (GFSK) modulation where each data bit is mapped into 1 of 2 frequencies. The 2 Mbps optional enhanced rate uses 4-level GFSK. In this case, 2 data bits are mapped into 1 of 4 frequencies. The filtered data is then modulated using standard frequency deviation. The BT value of 0.5 is chosen based on two competing factors—the need to achieve

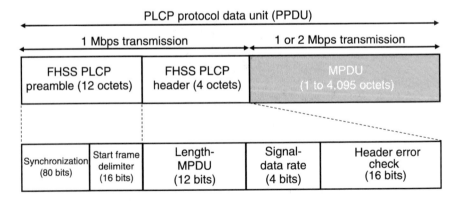

Figure 4.6 FHSS PLCP packet format.

bandwidth efficiency and the ability to tolerate intersymbol interference. High values for BT result in low intersymbol interference at the expense of high bandwidth requirements. Both 2-GFSK and 4-GFSK have the same root-mean-square (rms) deviation from the frequency carrier. Binary data is first filtered in the baseband using a low-pass Gaussian filter (500 KHz bandwidth) with a bandwidth-time product (BT) of 0.5. Table 4.2 shows the carrier frequency deviations for 2-GFSK and 4-GFSK. Figure 4.7 illustrates the operation of these modulation schemes.

Each frequency channel in a hopping pattern occupies 1 MHz of bandwidth and must hop at the minimum rate specified by regulatory bodies. For example, a minimum hop rate of 2.5 hop/s (corresponding to a maximum dwell time of 400 ms) is specified for the United States. The dwell time can be modified through the access points to suit certain propagation conditions. Once set, the dwell time remains constant. The dwell time is learned by a mobile node when it associates with the access point. This allows the node to remain synchronized to the access point while hopping between different frequency channels. The hopping patterns specified in the 802.11 standard (see Table 4.3) minimizes the probability of one BSS operating on the same frequency channel at the same time as another BSS. Sequences from the same set collide 3 times on the average (5 times worst case) over a hopping pattern cycle. In addition, the patterns are designed to ensure some minimum separation in frequency channels between contiguous hops. The separation provides some degree of diversity against frequency-selective multipath fading.

Table 4.2
Carrier Deviations for 2-GFSK and 4-GFSK

Modulation	**Data**	**Carrier Deviation**	**Modulation Index**
2-GFSK	0	$-0.5 \times h_2 \times$ Symbol Rate	0.160
	1	$+0.5 \times h_2 \times$ Symbol Rate	0.160
4-GFSK	00	$-1.5 \times h_4 \times$ Symbol Rate	0.216
	01*	$-0.5 \times h_4 \times$ Symbol Rate	0.072
	11	$+0.5 \times h_4 \times$ Symbol Rate	0.072
	10*	$+1.5 \times h_4 \times$ Symbol Rate	0.216
* Leftmost bit is transmitted first.			

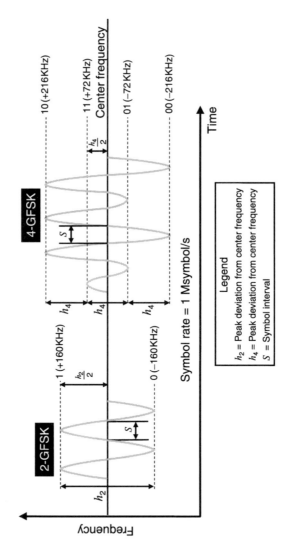

Figure 4.7 Multilevel GFSK modulations.

Table 4.3
Frequency Channels and Hopping Patterns of the FHSS 802.11 Standard

Country	Minimum Number of Frequency Channels	Actual Number of Frequency Channels	Number of Sets of Hopping Patterns	Number of Hopping Patterns in Each Set	Number of Hopping Patterns
USA	75	79	3	26	78
Most of Europe	20	79	3	26	78
Japan	10	23	3	4	12
Spain	20	27	3	9	27
France	20	35	3	11	33

The minimum hop distance is 6 MHz for the United States and Europe (including Spain and France) and 5 MHz for Japan.

The MPDU is scrambled and formatted to limit dc offset changes. The gradual increase (ramp-up) and decrease (ramp-down) of transmit power reduce splatter (spikes) in adjacent frequency channels at the start and end of each packet. Up to 8 μs has been allocated for the signal power to ramp-up to the desired level. Note that DSSS transmission requires much shorter ramp-up times (2 μs) due to a lower transmit power level.

4.2.4 Infrared physical layer

The DFIR PHY layer operates in the 850 to 950 nm wavelength using pulse position modulation (PPM) with a peak power of 2W. In general, a L-PPM system will break up the symbol period into L subintervals or time slots. A narrow pulse of infrared radiation is transmitted in one of the time slots. Thus, just like multilevel modulation, the symbol rate can be made slower than the data rate. However, unlike multilevel modulation, the bandwidth in L-PPM is increased by a factor of $L/\log_2 L$ relative to on-off intensity modulation. Thus, although more bits can be transmitted when time slots of shorter intervals are chosen, the narrower light pulses needed to fit into the slots lead to a higher bandwidth requirement. Extra noise introduced by the extra bandwidth may limit the performance of L-PPM.

The DFIR 802.11 PLCP packet format is shown in Figure 4.8. The first three fields are transmitted using on-off keying intensity modulation. The direct current level adjustment (DCLA) is used to allow the receiver to stabilize the average signal level after the transmission of the first three fields. The start frame delimiter (SFD) pattern requires careful selection because it directly affects the packet error rate. The probability that the SFD is correctly detected depends on the probability of imitation and the probability of error of the SFD [1]. The 802.11 standard adopted the pattern 1001 which is one of the patterns that maximizes the probability of error detection in the SFD field. The shadowed fields are transmitted using L-PPM. Note that the maximum length for the DFIR MPDU is the shortest compared to DSSS and FHSS.

The 1 Mbps DFIR standard uses PPM with 16 positions (16-PPM) where 4 data bits are mapped into 1 of 16 pulses (see Figure 4.9). The 2 Mbps version uses 4-PPM where 2 data bits are mapped into 1 of 4 pulses (see Figure 4.10). Regardless of the data rate supported, the duration of each L-PPM time slot has been specified to be 250 ns. This means that for 16-PPM, 4 bits of information can be transmitted in an interval of 4 ms (16 slots × 250 ns/slot), thus giving a wireless data rate of 1 Mbps. In a similar way, a 4-PPM infrared LAN transmits 8 data bits in an interval of 4 ms and this produces a data rate of 2 Mbps.

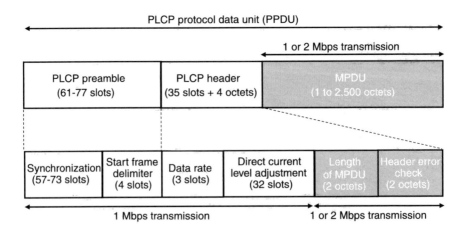

Figure 4.8 PLCP packet format for the IEEE 802.11 infrared LAN.

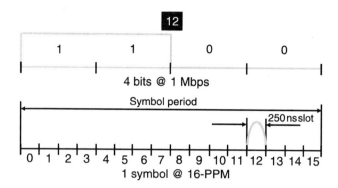

Figure 4.9   Pulse position modulation signals at 1 Mbps.

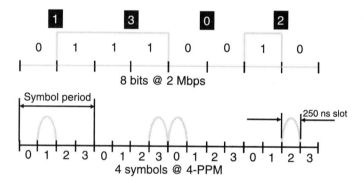

Figure 4.10   Pulse position modulation signals at 2 Mbps.

4.3   IEEE 802.11 medium access control layer

The 802.11 MAC layer is primarily concerned with the rules for accessing the shared wireless medium. Two different access methods have been defined. The function of the MAC protocol is common to all three PHY layer options (i.e., DSSS, FHSS, DFIR) and is independent of the data rates. The standard includes a formal description of the MAC protocol using the SDL method standardized by the International Telecommunications Union-Telecommunication Sector (ITU-T). The main services provided by the MAC layer are discussed in the following sections.

4.3.1 General 802.11 MAC protocol data unit

The general 802.11 MAC protocol data unit (MPDU) format is shown in Figure 4.11. The fields Address 2, Address 3, Sequence Control, Address 4 and User Data are only present in certain types of packets. The MPDU is separately protected by error checking bits. There are three types of packets namely:

1. Data packets;

2. Control packets (e.g., RTS, CTS, ACK packets);

3. Management packets (e.g., beacons).

The information provided by the various fields in the MPDU header are listed in Table 4.4. Some of the terms will be explained in the following sections.

4.3.2 Interframe spaces

Three different time gaps or interframe spaces (IFSs) are defined (see Figure 4.12). These interframe spaces are independent of the wireless data rate. The short IFS (SIFS) is the shortest IFS and is used for all immediate response actions (e.g., transmission of ACK, RTS, CTS packets). The point coordination function IFS (PIFS) is an intermediate-length IFS that is used for polling nodes with time-bounded requirements. The distributed coordination function IFS (DIFS) is the longest IFS that is used as a minimum delay between successive data packets. A slot time is also

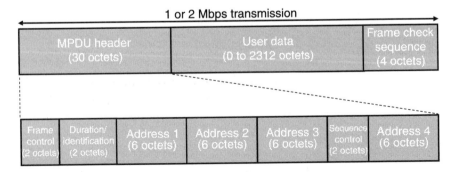

Figure 4.11 General MAC protocol data unit (MPDU) format.

Table 4.4
Information Provided by Various Fields of the MPDU Header

Field	Information Provided
Frame Control	Current version of standard, packets received or forwarded to the distribution system, power management, packet fragmentation, encryption and authentication.
Duration/Identification	Duration of net allocation vector, Identity of node operating in the power-save mode.
Addresses 1 to 4	Addresses for BSSID, destination, source, transmitter, and receiver.
Sequence Control	Sequence number for packet and packet fragment.

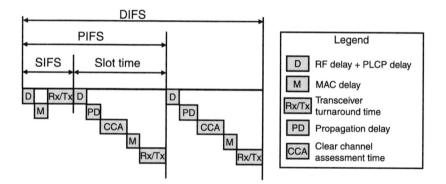

Figure 4.12 Interframe space definitions.

defined and this corresponds to a time slot used for backoff purposes (see Section 4.3.3). The slot time is the sum of the channel assessment (carrier sensing) time, the transceiver turnaround time, the signal propagation delay, and the MAC processing delay. SIFS is a function of the receiver delay, delay in decoding the PLCP preamble/header, the transceiver turnaround time, and the MAC processing delay. The 802.11 standard defines different values for the slot time and the SIFS for different physical layers. For example, in DSSS LANs, the 802.11 standard states that SIFS = 10 μs and slot time = 20 μs. For FHSS LANs, SIFS = 28 μs and slot time = 50 μs. DIFS is defined to be SIFS + 2 × slot time while the PIFS is specified as SIFS + slot time. As shown in Table 4.5, the IFSs for DSSS systems are at least two times shorter than FHSS systems. This means that

Table 4.5
Interframe Space Specifications

Interframe Space	DSSS	FHSS	DFIR
SIFS	10 µs	28 µs	7 µs
PIFS	30 µs	78 µs	15 µs
DIFS	50 µs	128 µs	23 µs
Slot time	20 µs	50 µs	8 µs

a DSSS transmission incurs less overhead due to the interframe time gaps. Note that the slot time for 10 Mbps Ethernet is defined as 512 bit times or 51.2 µs. However, this duration also takes into account the time needed for collision detection.

4.3.3 Distributed coordination function

The basic access method in the 802.11 standard is called the distributed coordination function (DCF) which is essentially carrier sense multiple access with collision avoidance (CSMA/CA). CSMA/CA is very similar in operation to the carrier sense multiple access with collision detection (CSMA/CD) protocol employed by wired Ethernet networks. In both protocols, the availability of the transmission medium is detected through carrier sensing, and contention on the medium is resolved using an exponential backoff algorithm. Since the CSMA/CA protocol uses distributed control, there is no central controller that coordinates access to the wireless LAN. Thus, nodes can transmit when they wish to, as long as they obey the protocol rules.

Carrier Sense Multiple Access

In CSMA, a node with a packet to transmit first senses the wireless medium to detect any ongoing transmission. If the medium is busy (i.e., some other node is transmitting), the node defers its transmission to a later time. If the medium is sensed to be free for a duration greater than the DCF interframe space (DIFS) interval, the packet is transmitted immediately. The MAC layer operates in conjunction with the PHY layer to assess the condition of the medium. One method involves measuring the radio signal energy to determine the strength of the received signal.

If the received signal strength is below a specified threshold, the medium is declared clear and the MAC layer is given the clear channel assessment (CCA) status for packet transmission. Another method correlates the received signal with the 11-chip Barker code to detect the presence of a valid DSSS signal. Both methods may also be combined to provide a more reliable indication of the status of the medium.

CSMA is very effective when the medium is lightly loaded since the protocol allows nodes to transmit with minimum delay. Due to a finite propagation delay along the transmission medium, there is a probability of two or more nodes simultaneously sensing the medium as being free and transmitting at the same time, thereby causing a collision. Clearly, such collisions will occur often when the network is heavily loaded with many transmitting nodes. The ratio of the slot time over the packet transmission time also affects the performance of CSMA. As shown in Table 4.6, the ratio of the slot time (defined in Table 4.5) over a standard Ethernet packet is reasonably small for the CSMA algorithm in the 802.11 standard to operate efficiently. At higher data rates, CSMA may not operate efficiently for short Ethernet packets.

Collision Avoidance

The CSMA protocol incorporates a collision avoidance (CA) scheme that introduces a random interframe space (backoff) between successive packet transmission. Collision avoidance is performed to reduce the high probability of collision immediately after a successful packet transmission. It is essentially an attempt to separate the total number of transmitting nodes into smaller groups, each using a different time slot (known as a backoff time slot). If the medium is detected to be busy, the node must

Table 4.6
Ratio of the Slot Time over Different Ethernet Packet Lengths
(Wireless Overhead Ignored)

Ethernet Packet	DSSS		FHSS		DFIR	
Length (Octets)	1 Mbps	2 Mbps	1 Mbps	2 Mbps	1Mbps	2Mbps
1518	0.0016	0.0033	0.004	0.008	0.0007	0.0013
512	0.005	0.010	0.012	0.024	0.002	0.004
64	0.039	0.078	0.098	0.195	0.016	0.031

first delay until the end of the DIFS interval and further wait for a random number of time slots (called the backoff interval) before attempting transmission (see Figure 4.13). When retransmission is necessary, the backoff interval increases exponentially up to a certain threshold. Conversely, the backoff interval reduces to a minimum value when packets are transmitted successfully. This is how backoff intervals of random lengths are used to resolve conflicts due to collisions.

At each backoff time slot, carrier sensing is performed to determine if there is activity on the medium. If the medium is idle for the duration of the slot, the backoff interval is decremented by one time slot. If a busy medium is detected, the backoff procedure is suspended and the backoff timer will not decrement for that slot. In this case, when the medium becomes idle again for a period greater than the DIFS, the backoff procedure continues decrementing from the time slot which was previously disrupted. This implies that the selected backoff interval is now less than the first. Hence, a packet that was delayed while performing the backoff procedure has a higher probability of being transmitted earlier than a newly arrived packet. The process is repeated until the backoff interval reaches zero and the packet is transmitted.

The collision avoidance mechanism also ensures some degree of fairness since it forces a node with multiple packets to undergo the backoff procedure, thereby allowing another node a chance to transmit (see Figures 4.14 and 4.15). This mechanism is not used when a node decides to transmit a new packet and the medium has been free for a duration greater than the DIFS.

Collision and Error Detection

The collision detection mechanism used on a wired LAN requires the receiver to sense the medium while transmitting. This method cannot be directly applied to a wireless LAN for several reasons. First, in wired networks, the difference between the transmit and receive signal levels (i.e., the dynamic range) is small enough to detect collisions. However, in a wireless environment, the transmitted signal energy radiates in all directions and receivers typically have to be very sensitive to detect the signal. Since the receiver is colocated with the transmitter, this means that even when two or more nodes transmit at the same time, such collisions will be hard to detect because the transmission from a sending

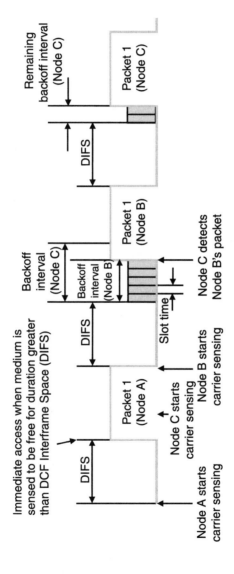

Figure 4.13 Single packet transmission using CSMA/CA.

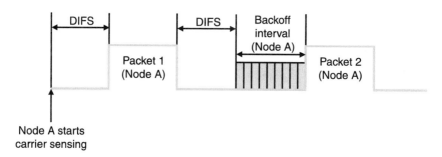

Figure 4.14   Multiple packet transmission using CSMA/CA
               (single node).

node will overpower transmission from all other nodes. Furthermore, the
basic assumption of collision detection requires all nodes to be able to hear
each other. This may not be practical in a wireless environment because
the high and variable attenuation of the signal makes detection of
colliding packets difficult. This is made worse by a hidden node situation
when a transmitting node senses the medium as free but the medium in
the region surrounding the receiver is not. Finally, collision detection
requires the implementation of costly two-way, full-duplex radio transceivers capable of transmitting and receiving at the same time.

The 802.11 MAC protocol requires the receiver to send a positive
acknowledgment (ACK) back to the transmitter if a packet is received
correctly (see Figure 4.16). The ACK is transmitted after the short
interframe space (SIFS) which is of a shorter duration than the DIFS. This
enables an ACK to be transmitted before any new packet transmission.
If no ACK is returned, the transmitter assumes the packet is corrupted
(either due to a collision or transmission error) and retransmits the
packet. Hence, unlike CSMA/CD, in CSMA/CA collisions are inferred
only after the entire data packet is transmitted. Since retransmission is
performed by the MAC layer and not by higher layers, this ensures fast
recovery of lost messages. For wireless LANs, implementing error recovery at the MAC layer is also more efficient since errors occur more
frequently than wired networks. On the other hand, the use of a positive
acknowledgment lowers the transmission efficiency since every correctly
received packet has to be acknowledged. The 802.11 standard requires
ACKs to be issued only on receipt of unicast packets. Acknowledgments
for broadcast or multicast traffic are not practical since this will lead to

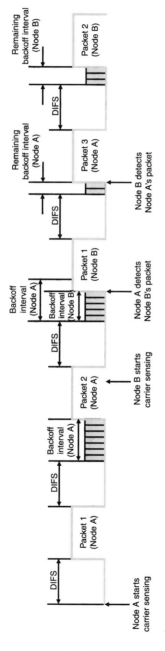

Figure 4.15 Multiple packet transmission using CSMA/CA (multiple nodes).

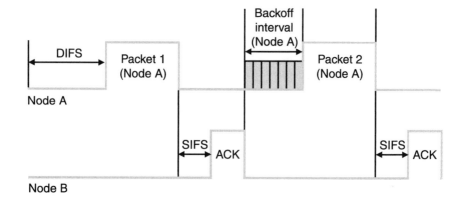

Figure 4.16 Successful transmission of a unicast data packet.

collisions among multiple ACKs. As a result, reliability for such traffic is reduced. Note that without collision detection, packets transmitted using CSMA need not be of a certain minimum length.

Virtual Sensing

CSMA/CA can be enhanced by incorporating the virtual carrier sense mechanism which distributes reservation information by announcing the impending use of the wireless medium. The exchange of short control packets called request-to-send (RTS) and clear-to-send (CTS) packets prior to the transmission of the data packet serves this purpose (see Figure 4.17). The RTS packet is issued by the transmitting node while the CTS packet is issued by the receiving node to grant the requesting node permission to transmit. The RTS and CTS packets contain a duration field that defines the period of time the medium is to be reserved for the transmission of the data packet and the returning ACK packet. The short RTS and CTS packets minimize the overhead due to collisions and also allow the transmitting node to infer collisions quickly. In addition, the CTS packet alerts neighboring network nodes (that are within the range of the receiving node but not of the transmitting node) to refrain from transmitting to the receiving node, thereby reducing hidden node collisions (see Figure 4.18). In the same way, the RTS packet protects the transmitting area from collisions when the ACK packet is sent from the receiving node (see Figure 4.19). Thus, reservation information is distributed around both the transmitting and receiving nodes. All other nodes

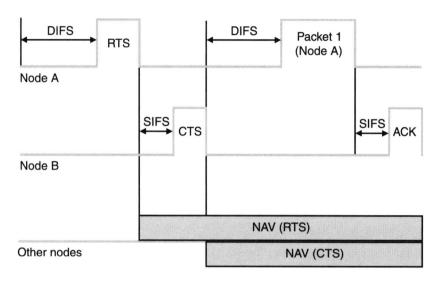

Figure 4.17 Unicast data packet transmission using virtual sensing.

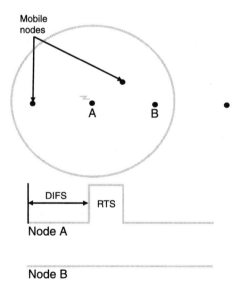

Figure 4.18 Transmission of the RTS packet.

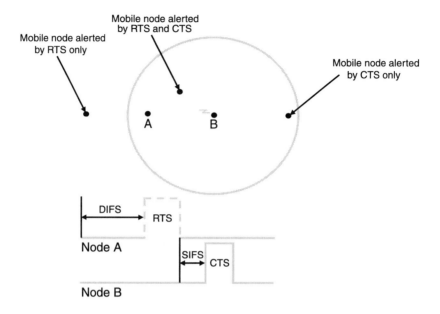

Figure 4.19   Transmission of the CTS packet.

that successfully decode the duration field in the RTS and CTS packets store the medium reservation information in a net allocation vector (NAV). For these nodes, the NAV is used in conjunction with carrier sensing to detect the availability of the medium. Hence, these nodes will defer transmission if the NAV is nonzero or if carrier sensing indicates that the medium is busy. Like the ACK mechanism, virtual sensing cannot be applied to MPDUs with broadcast and multicast addressing due to a high probability of collision among a potentially large number of CTS packets. Because of the large overhead involved, the mechanism is not always justified, particularly for short data packets. Hence, the 802.11 standard allows short packets to be transmitted without virtual sensing. This is controlled by a parameter called the RTS threshold. Only packets of lengths above the RTS threshold are transmitted using virtual sensing. Note that the effectiveness of the virtual sense algorithm depends strongly on the assumption that both the transmitting and receiving nodes have similar operating ranges (i.e., transmitter power and receiver sensitivity are about the same). Use of virtual sensing is optional but the mechanism must be implemented.

4.3.4 Point coordination function

Real-time traffic requires bounded end-to-end delays beyond which the information loses its value and may be discarded. This is in contrast to the delay requirements for data traffic which are less stringent. CSMA/CA is not particularly suited for real time traffic support because it treats all packets equally without taking into account the sensitivity of certain types of data. Its connectionless nature does not schedule or prioritize time-sensitive real-time traffic (e.g., voice, video) and as a result, is unable to distinguish between such traffic and nonreal-time traffic (e.g., data). The possibility of collisions, the use of a random backoff interval and the transmission of long packets may also lead to excessive delay variation (jitter). A further point to note is that the use of a positive acknowledgment for collision and error detection in CSMA/CA may degrade the transmission of real-time traffic since retransmission increases the delay.

There is an optional point coordination function (PCF) which may be used to support time-bounded services. PCF employs a centralized, contention-free multiple access scheme where nodes are allowed to transmit only when polled by the access point. Note that collisions may be introduced when access points transmit polling messages to mobile nodes stationed in overlapped wireless coverage areas. To allow other nodes with asynchronous data to access the medium, the MAC protocol alternates between DCF and PCF, with PCF having higher priority access. This is achieved using a superframe concept where the PCF is active in the contention-free period, while the DCF is active in the contention period (see Figure 4.20). The contention-free period can be variable in length within each superframe without incurring any additional overhead. At

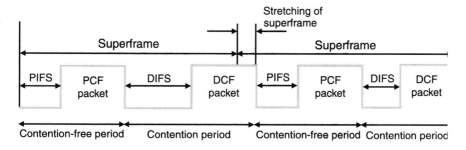

Figure 4.20 Coexistence of PCF and DCF in a superframe.

the beginning of the superframe, if the medium is free, the PCF gains control over the medium. If the medium is busy, then PCF defers until the end of the packet or after an acknowledgment has been received. Since the PIFS is of shorter duration compared to DIFS (see Figure 4.12), PCF can gain control of the medium immediately after the completion of a busy period. Since the contention period may be of variable lengths, this causes the contention-free period to start at different times (see Figure 4.20). Similarly, a packet may start near the end of the contention period, thereby stretching the superframe and causing the contention-free period to start at different times.

4.3.5 Association and reassociation

Association enables the establishment of wireless links between mobile nodes and access points in infrastructure networks. A node joins a network and is capable of transmitting and receiving data packets only after the association process is completed. To initiate a new connection with an access point, the node transmits a Probe signal. On receiving a Probe Response from the access points, the node selects the access point with the best signal strength. The node then sends an Association Request to the access point which will issue an Association Response.

Association is necessary but not sufficient to support mobility. In order to support mobility and roaming, an additional function called reassociation must be used in conjunction with the association function. Reassociation enables an established association to be transferred from one access point to another. Reassociation is always initiated by the mobile node. Association and reassociation are dynamic processes since mobile nodes may power on, power off, move within range, or go out of range.

To illustrate how both processes interact in a roaming situation, consider the case when a node decides that a link to its current access point is poor. The node scans for another access point or uses information from previous scans. If a new access point is located, the node sends a Re-association Request to a new access point. If the Reassociation Response is successful, the node becomes connected to the new access point. Otherwise, the node scans for another access point. When the access point accepts a Reassociation Request, it indicates Reassociation to the Distribution System (DS). Information on the DS is then updated and the old access point is normally notified of the change through the DS.

4.3.6 Authentication and privacy

By default, messages are transferred across 802.11 wireless LANs without encryption. Hence, a DSSS 802.11-compliant node can potentially eavesdrop DSSS 802.11 traffic that is within range. The 802.11 standard includes an optional provision for security called wired equivalent privacy (WEP). The WEP option uses a complex 40-bit encryption and authentication stream cipher (based on RSA's RC4) algorithm to encrypt data before transmission. Implementation of the WEP algorithm requires the use of a pseudorandom generator that is initialized by a secret key. This generator outputs a key sequence equal to the largest possible packet length. It is then combined with the user data to produce the packet to be transmitted over the wireless medium. In order to operate in a connectionless LAN environment, every packet is sent with a 24-bit initialization vector which restarts the pseudorandom generator for each packet. A common 64-bit shared key is used to authenticate, encrypt, and decrypt the data. Only those devices with a valid shared key will be allowed to be associated to the access point. The WEP only protects the MPDU information and not the PHY layer header.

The basic procedures for authentication (used in conjunction with the encryption features of WEP) are as follows:

▶ Node A challenges the identity of node B by transmitting a random message;

▶ Node B encrypts the message using WEP and sends it back to node A;

▶ Node A decrypts message and authentication is successful if the decrypted message matches the original message; otherwise, Node A issues a new challenge using a different random message.

4.3.7 Synchronization

Mobile nodes need to keep synchronization for several reasons (e.g., chipping rate/hopping rate synchronization and power management).

For accuracy, the timing synchronization function requires two items of information at each node: a common clock reference and the forward/ reverse propagation times. The latter allows each node to compensate for the differences in distance from other nodes. Beacons play an important role in network synchronization since they convey important network parameters such as hop sequence and timing information. Beacons also enable new mobile nodes to join a network.

In an infrastructure network, synchronization can be achieved using the following mechanism:

▶ The access point periodically transmits beacons which contain the timing information of the access point at the instant of transmission;

▶ The receiving nodes adjust their local clocks at the moment the beacon signal is received.

When a node that wishes to access an existing BSS (either after power-up, power-save mode, or entering a BSS), the node can obtain synchronization information through the following ways:

▶ Passive scanning where a node waits to receive a beacon from the access point;

▶ Active scanning where a node tries to locate an access point by transmitting Probe Request packets and then listens for a Probe Response from the access point.

Synchronization becomes a distributed function for IBSS networks.

4.3.8 Power management

Power conservation is critical for wireless LANs since mobile devices are battery-powered. LAN adapter receiver circuits are typically active for a longer period of time than transmitter circuits. However, when averaged over time, the idle receive state usually dominates LAN adapter power consumption. Hence, considerable battery power can be saved by allowing the node to switch off during idle periods and yet maintain an active connection.

The 802.11 standard defines three power modes that are incorporated into the MAC protocol:

1. Transmit: Transmitter is activated;

2. Awake: Receiver is activated;

3. Doze: Transceiver is not able to transmit or receive.

The 802.11 MAC protocol allows the mobile node to switch from full power (active) mode to the low power (sleep) mode during a time interval defined by the access point without losing information. Actual power consumption is not defined in the standard and is implementation-dependent. The main principle behind the power saving mechanism is that the access point keeps an updated record of the nodes currently operating in the power saving mode. It then buffers the packets addressed to these nodes until the nodes specifically request for the packets or when they resume communication with the access point. If the node moves to a different access point, packets are forwarded across the wired LAN to the node.

The access point periodically sends beacons with the traffic indication map (TIM) that announces which power saving nodes have unicast traffic stored (see Figure 4.21). The beacon transmission may be delayed by an ongoing data transmission. In the power-save mode, the mobile node

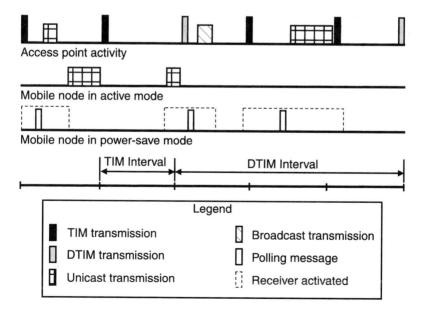

Figure 4.21   Power management in IEEE 802.11 wireless LANs.

regularly wakes up prior to any TIM broadcasts from an access point. However, the node need not check every TIM broadcast. If there are packets intended for the node, it shifts from doze mode to awake mode and sends a polling message to the access point to retrieve these packets. In order to maintain synchronization, a timer continues to run as a node sleeps. Synchronization allows extreme low-power operation. The availability of broadcast, multicast, and unicast messages is indicated via the delivery TIM (DTIM). Broadcast and multicast messages are sent first. The DTIM interval has a longer duration and is a multiple of the TIM interval. For an IBSS, since no access point is available, beacon transmission becomes a distributed responsibility. Adhoc TIMs are issued before the actual data packet is transmitted. The power-conserving nodes wake up only for a short predefined period of time to hear if they have to remain active to receive a packet.

4.3.9 Packet fragmentation

As in most LAN packet transmission, variable message lengths are employed in the IEEE 802.11 standard. In this way, the total number of packets sent is minimized. This can be important in obtaining high throughput since many network devices are limited not by the number of bits they can transmit per second but by the number of packets they can process per second [2]. This is especially true for large coverage areas since the bit error rate of the wireless medium increases with distance. The fragmentation feature can also be useful when applied to mobile devices that travel at medium velocity (e.g., a device on a forklift truck moving in a warehouse at 15 km/h). Normally, fading occurs rapidly under such conditions. Packet fragmentation can lessen the impact of collisions and is a good alternative to the use of RTS/CTS (although the 802.11 standard allows fragmentation to be used in conjunction with the RTS/CTS mechanism). The 802.11 standard recommends a fragmentation length of duration less than 3.5 ms (i.e., a packet length of 400 octets for a data rate of 1 Mbps) [3]. However, fragmentation inherently requires more overhead due to an increased number of packets/ACKs processed, the need for preamble/header information in each fragmented packet, and the need for extra SIFSs.

To achieve these benefits, a simple fragmentation/reassembly mechanism is incorporated into the 802.11 MAC layer (see Figure 4.22). Each

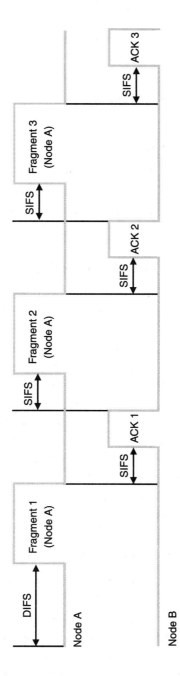

Figure 4.22 Fragmentation of a unicast data packet.

packet contains a sequence number for reassembly purposes. A frag-mented threshold determines the maximum length of a packet above which the packet will be fragmented.

4.4 Unresolved issues of the IEEE 802.11 standard

Standardization and interoperability among devices utilizing the same PHY layer is the intent of the 802.11 specification. However, even among devices with the same PHY, some key issues necessary to achieve multi-vendor interoperability are absent in the ratified standard.

4.4.1 Access point coordination for roaming

The 802.11 standard includes several provisions necessary for roaming but it does not specify a particular algorithm. For example, it does not specify the handoff mechanism to allow mobile nodes to roam from one access point to another. Thus, several roaming options can be applied. There have been efforts to define an inter-access point protocol (IAPP) that facilitates roaming across access points from multiple vendors. The IAPP also deals with security issues and caters to wireless infrastructure distribution systems that have no wired backbone. Although stan-dardization of IAPP has not been pursued further, the specification is available (in form of an Internet-draft) and several vendors have imple-mented it or are in the process of doing so. Currently, no interoperability testing of IAPP has taken place.

4.4.2 Conformance test suite

There is no conformance test suite specified to verify that a device is compliant with the 802.11 specification. Vendor claims for compliance to the 802.11 standard will need to be ratified by a neutral (independent) third party. The University of New Hampshire is currently the preferred conformance test facility that verifies compliance to the 802.11 specifica-tion and performs multivendor interoperability testing.

4.5 Current developments of the IEEE 802.11 standard

The 802.11 standard committee is now focusing on developing higher speed standards in the 2.4 GHz and 5 GHz frequency bands. The 2.4 GHz PHY layer is targeted to support a wireless data rate of 11 Mbps and is an extension of the existing 1 and 2 Mbps 802.11 DSSS radio standard. The high-speed proposal supports fallback rates of 5.5, 2, and 1 Mbps. It is expected that the standard will be formally approved by early 2000. Further information about 802.11 activities can be obtained from http://grouper.ieee.org/groups/802/11/index.html.

4.5.1 The high-speed 2.4 GHz standard

The 11 Mbps technology employs DQPSK modulation in conjunction with an 8-chip spreading code known as complementary code keying (CCK). This CCK scheme is based on orthogonal subsets of codes that are generated from the sequence [+1, +1, +1, −1, +1, +1, −1, +1]. In order to retain the complementary characteristics of this sequence, phase components are added as follows:

$$
8-\text{Chip Complex Code } (\phi_1, \phi_2, \phi_3, \phi_4) =
$$
$$
\left[e^{j(\phi_1 + \phi_2 + \phi_3 + \phi_4)}, e^{j(\phi_1 + \phi_3 + \phi_4)}, e^{j(\phi_1 + \phi_2 + \phi_4)}, -e^{j(\phi_1 + \phi_4)}, \right.
$$
$$
\left. e^{j(\phi_1 + \phi_2 + \phi_3)}, e^{j(\phi_1 + \phi_3)}, -e^{j(\phi_1 + \phi_2)}, e^{j\phi_1} \right]
$$
$$
\text{where } \phi_i \in \left\{ 0, \frac{\pi}{2}, \pi, \frac{3\pi}{2} \right\}
$$

The phases are based on generalized Hadamard encoding. In this coding scheme, for a code of length-N (where $N = 2^n$), $n + 1$ phases are encoded into 2^n output phases by adding the first phase to all code phases, the second to all odd code phases, the third to all odd pairs of code phases, and so forth. For example, a length-8 code can be written as

$$
\begin{bmatrix} \theta_1 \\ \theta_2 \\ \theta_3 \\ \theta_4 \\ \theta_5 \\ \theta_6 \\ \theta_7 \\ \theta_8 \end{bmatrix} = \begin{bmatrix} 1111 \\ 1011 \\ 1101 \\ 1001 \\ 1110 \\ 1010 \\ 1100 \\ 1000 \end{bmatrix} \begin{bmatrix} \phi_1 \\ \phi_2 \\ \phi_3 \\ \phi_4 \end{bmatrix}
$$

Since ϕ_1 is common in all 8 chips, the entire complex code is rotated by ϕ_1. Assuming that ϕ_1 can take on any of the 4 phases (i.e., 0, $\pi/2$, π, $3\pi/2$), simulation studies indicate that the certain values for ϕ_2, ϕ_3, and ϕ_4 can tolerate large multipath delay spreads. These values also help to minimize the number of correlators needed for implementing CCK. The values are $\phi_2 = (\pi/2, 3\pi/2)$, $\phi_3 = 0$, and $\phi_4 = (0, \pi)$. Substituting these values yields the following 8-chip complex code:

$$
8-\text{Chip Complex Code } (\phi_1, \phi_2, \phi_3, \phi_4) =
$$
$$
e^{j\phi_1} \begin{bmatrix} e^{j(\phi_2 + \phi_4)}, e^{j(\phi_4)}, e^{j(\phi_2 + \phi_4)}, -e^{j\phi_4}, \\ e^{j\phi_2}, -e^{j\phi_2}, 1 \end{bmatrix}
$$
$$
\text{where } \phi_1 \in \left\{ 0, \frac{\pi}{2}, \pi, \frac{3\pi}{2} \right\}, \phi_2 \in \left\{ \frac{\pi}{2}, \frac{3\pi}{2} \right\}, \phi_4 \in \{0, \pi\}
$$

A complementary pair of codes that can be generated from this complex code is illustrated in Figure 4.23.

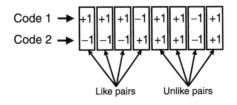

Figure 4.23   Complementary code pair.

In the proposed 5.5 Mbps and 11 Mbps standards, the 8-CCK scheme employs the same chipping rate of 11 Mchip/s (and hence the same spreading bandwidth of about 22 MHz) as the current 802.11 standard for DSSS transmission. As such, it is backward-compatible with current 1 and 2 Mbps DSSS 802.11 devices. Since the chip length is reduced from 11 to 8, the symbol rate has to be increased to 1.375 Msymbol/s in order to maintain the 11 Mchip/s chipping rate (see Figure 4.24). In the 11 Mbps CCK, 6 bits are used to select 1 of 64 complex codes (of 8-chip length) while the remaining 2 bits are used to modulate the entire code using DQPSK. For the 5.5 Mbps version, 2 bits select 1 of 4 codes while the other 2 bits modulate the entire vector using DQPSK. Matched filters may be sufficient for implementing the CCK scheme (i.e., sophisticated equalizers are not necessary). Although CCK is found to be resistant to the delay spread caused by multipath propagation (100 ns at 11 Mbps and 250 ns at 5.5 Mbps), the new 11 Mbps proposal operates with a reduced range. Note that CCK possesses both processing and coding gains which combine to produce an overall gain that meets the minimum 10 dB requirement for 2.4 GHz DSSS systems.

4.5.2 The high-speed 5 GHz standard

The IEEE 802.11 standard committee is also developing a 5 GHz PHY layer that will support wireless data rates of 6 to 54 Mbps. Activities in the United States, Europe, and Japan suggest that radio spectrum in the

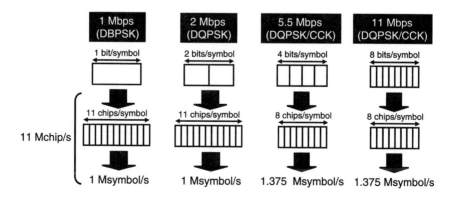

Figure 4.24 Data rates for the IEEE 802.11 standard.

5 GHz range may follow the direction of the 2.4 GHz band and become available worldwide in due course. The 802.11 working group has organized a joint meeting with the relevant project group of ETSI to establish cooperation on the new high-speed PHY layer based on orthogonal frequency division multiplexing (OFDM) with convolutional coding. Chapter 6 will cover OFDM and convolutional coding in more detail. The main specifications are listed in Table 4.7. The 802.11 will employ its own MAC protocol, which has the distributed access control based on asynchronous packet transfer while ETSI BRAN is working on a MAC protocol that is centralized and is based on ATM. ETSI BRAN has a liaison with the ATM Forum—details are provided in Chapter 7. The 5 GHz devices will require more transmit power to achieve the same range as 2.4 GHz adapters.

4.5.3   The personal area network study group

The 802.11 committee is also developing standards for networking wearable devices within a very small area. The Personal Area Network (PAN) study group hopes to establish a standard that allows low-power devices (with the associated software) to communicate while being worn or carried by individual users. A minimum of 16 devices is required for each network.

Table 4.7
Specifications for the 5 GHz IEEE 802.11 Physical Layer

Parameters	Specification
Mandatory Data Rates (Mbps)	6, 12, 24
Optional Data Rates (Mbps)	18, 36, 48, 54
Number of Subcarriers	52 (48 for data, 4 for pilots)
Channel Spacing	20 MHz
Signal Bandwidth	16.6 MHz
Modulation for Subcarrier	BPSK, QPSK, 16-QAM, 64-QAM
Bit-Interleaved Convolutional Coding	$K = 7, R = 1/2, 2/3, 3/4$

4.6 Commercial IEEE 802.11 wireless LANs

Some of the typical specifications for commercial 802.11 compliant wireless LANs are listed in Tables 4.8 and 4.9. Chip sets that implement the standard are also available. These chip sets provide a low-cost alternative to build 802.11 wireless LANs for customized applications. Complexity in circuit design is reduced considerably through a high

Table 4.8
DSSS IEEE 802.11 Wireless LANs

Parameter	Specification
Transmit Radio Power	35 mW (15 dBm)
Maximum Operating Range	90 m (semi-open environment) to 550 m (open environment)
Power Consumption	15 mA (sleep), 250 mA (receive), 300 mA (transmit)
Receiver Sensitivity (Bit Error Rate = 10^{-5})	−91 dBm to −94 dBm
Delay Spread (Packet Error Rate <1%)	400 to 500 ns
Bit Error Rate	Better than 10^{-8}
Number of Nodes	60 to 200

Table 4.9
FHSS IEEE 802.11 Wireless LANs

Parameter	Specification
Transmit Radio Power	10 mW (10 dBm) to 500 mW (27 dBm)
Maximum Operating Range	120 m (semi-open environment) to 1000 m (open environment)
Power Consumption	10 mA (sleep), 280 mA (receive), 490 mA to 680 mA (transmit)
Receiver Sensitivity (Bit Error Rate = 10^{-5})	−79 dBm to −85 dBm
Hopping Rate	10 hop/s
Number of Frequency Hops	Programmable to meet regulations of all countries.
Dwell Time	20 to 128 ms
Number of Co-located Coverage Areas	Up to 15

degree of component integration. The chip sets address many difficult wireless hardware design issues such as impedance matching and biasing compatibility. More information on the chip sets can be found at http://www.intersil.com.

4.7 The HIPERLAN
Type 1 standard

HIPERLAN Type 1 [4] is a wireless LAN standard that is ISO 8802 compatible (equivalent to the IEEE 802). Like the 802.11 standard, HIPERLAN Type 1 caters to both independent and infrastructure networks. However, it has only one physical layer specification which is not based on spread spectrum transmission. HIPERLAN Type 1 operates in the 5.15 to 5.30 GHz band (which is divided into five channels) with a peak power level of 1W. It supports low-mobility users (1.4 m/s) that carry asynchronous or isochronous traffic at a range of up to 50 m and a maximum wireless data rate of about 23.5 Mbps. The HIPERLAN Type 1 reference model is shown in Figure 4.25. It comprises a physical layer (PHY), a medium access control (MAC) sublayer, and a channel access control (CAC) sublayer. The physical layer protocol specifies the techniques for transmission, reception, and channel assessment. The CAC sublayer determines which nodes have the right to transmit. It defines a common service over a single wireless communication link and allows the access priority to be specified. The MAC sublayer conforms to the ISO MAC service definitions and includes provisions for power conservation

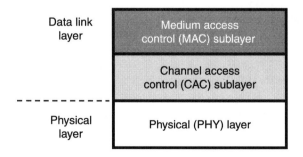

Figure 4.25 HIPERLAN Type 1 reference model.

and packet forwarding. ETSI standards and reports can be downloaded from http://webapp.etsi.org/publicationssearch.

4.7.1 Physical layer

HIPERLAN Type 1 employs the nondifferential Gaussian minimum shift keying (GMSK) modulation technique with a decision feedback equalizer (DFE) and a normalized filter bandwidth-time (BT) product of 0.3. GMSK is a constant envelope modulation scheme. This means that there is no variation in amplitude which allows the use of highly efficient power amplifiers. Nonconstant envelope modulation requires a high degree of linearity in power amplification. Failure to do so may result in unwanted intermodulation interference. However, the GMSK modulation scheme may require significant equalization overhead and high power consumption. HIPERLAN Type 1 supports 3 types of transmitter and receiver classes (Table 4.10). The receiver sensitivity is defined as the minimum power level that results in a packet error rate of 0.01 for a packet length of 4,160 bits. Clearly, a higher receiver sensitivity is required for HIPER-LAN Type 1 equipment operating at high transmit power.

Carrier sensing (clear channel assessment) on HIPERLAN Type 1 is based entirely on the received signal strength. A threshold is used to determine whether the medium is busy or idle. Error control is based on forward error correction using Bose-Chaudhuri-Hocquenghem codes or BCH (31, 26) as well as explicit acknowledgments. The BCH (31, 26) coding maps 26 data bits into 31 coded bits. Because each data block is interleaved across 16 codewords, this leads to a data block of 416 bits (52 octets) encoded to 496 (62 octets) bits. The coding scheme offers protection of at least 2 random errors and burst errors less than 32 bits long. A

Table 4.10
Combinations of Transmitter and Receiver Classes

	Transmitter Class A (+10 dBm)	Transmitter Class B (+20 dBm)	Transmitter Class C (+30 dBm)
Receiver Class A (–50 dBm)	Permissible	Not Permissible	Not Permissible
Receiver Class B (–60 dBm)	Permissible	Permissible	Not Permissible
Receiver Class C (–70 dBm)	Permissible	Permissible	Permissible

fuller discussion on error correction coding can be found in Section 6.3.12. The data packet format is shown in Figure 4.26. The different shades in each field correspond to the different layers of Figure 4.25.

Like the 802.11 counterpart, HIPERLAN Type 1 packets operate at 2 different wireless data rates. The low data rate operates at 1.4705875 Mbps (using FSK) while the 23.5294 Mbps data rate (using GMSK) is 16 times higher than the low rate. The low rate header contains enough information to inform a node whether it needs to listen to the rest of the packet or not. Thus, a node can keep the error correction, equalization, and other circuits powered off while listening unless the low rate header informs otherwise. The bandwidth required for high rate transmission is 23.5294 MHz. User data, acknowledgment, and synchronization information are transmitted at the high data rate. A minimum length is required for synchronization to ensure that the combined false alarm and misdetection rate is kept low. The transmitted data is coded as a number of fixed-length blocks (62 octets).

4.7.2 Comparison of IEEE 802.11 and HIPERLAN specifications

Table 4.11 compares the various specifications of the IEEE 802.11 and HIPERLAN standards. While the 802.11 standard provides better throughput efficiency, the low data rate results in longer delays when compared to HIPERLAN. Note that the wireless throughput efficiencies represent the upper bounds achievable since additional overhead needed

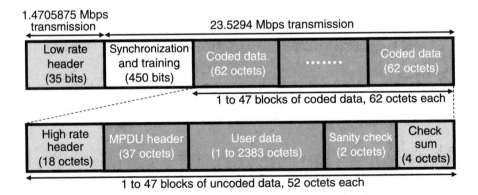

Figure 4.26 HIPERLAN Type 1 data packet format.

Table 4.11
Comparison of the IEEE 802.11 and HIPERLAN Specifications

Parameter	IEEE 802.11 DSSS	IEEE 802.11 FHSS	HIPERLAN
Idle time needed for immediate packet transmission	50 µs	128 µs	85 µs
Interframe space for transmission of acknowledgment	10 µs	28 µs	21.8 µs
Overhead for low data rate transmission	192 bits (1 Mbps)	128 bits (1 Mbps)	35 bits (1.47 Mbps)
Maximum wireless throughput efficiency for transmission of one packet	97.7% (2 Mbps), 98.8% (1 Mbps)	99.2% (2 Mbps), 99.6% (1 Mbps)	78.4%
Wireless throughput efficiency for transmission of one 1518-octet Ethernet packet	96.9% (2 Mbps), 98.4% (1 Mbps)	97.9% (2 Mbps), 99.0% (1 Mbps)	74.1%
Transmission time for one 1518-octet Ethernet packet	6,264 µs (2 Mbps), 12,336 µs (1 Mbps)	6,200 µs (2 Mbps), 12,272 µs (1 Mbps)	696.4 µs
Transmission time for acknowledgment	248 µs (2 Mbps), 304 µs (1 Mbps)	184 µs (2 Mbps), 240 µs (1 Mbps)	15.6 µs

for medium access (e.g., carrier sensing, contention, backoff) have been omitted in the calculations.

4.8 HIPERLAN Type 1 medium access control layer

HIPERLAN Type 1 employs a fully distributed MAC protocol called Elimination Yield Non Pre-emptive Multiple Access (EY-NPMA). EY-NPMA is essentially CSMA with added prioritization. Channel access is non pre-emptive because only data packets ready at the start of each channel access cycle are allowed to contend. Nodes undergo contention resolution based on priority assertion and random backoff (see Figure 4.27). Clearly, if two nodes have different access patterns, then the

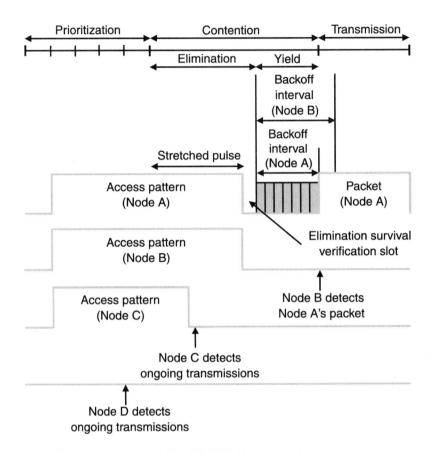

Figure 4.27 Operation of the EY-NPMA protocol.

listening and transmission periods as well as the eventual packet transmission will not coincide. The slot time is of variable length, depending on whether the slot is a listen period or a signal period. In the following discussion, all bit times are referenced to the higher data rate of 23.5294 Mbps.

The prioritization phase comprises 5 slots of 168 bits each. The packet with the highest priority to transmit corresponds to the packet with holding an access pattern with the largest decimal number. Hence, an access pattern with the highest priority level (i.e., 5) will show no idle slots at the beginning while an access pattern of the lowest priority will have 4 idle slots. The contention interval can be divided into two phases: elimination and yield. The elimination phase involves stretching the pulse with a random number of slots (0 to 12) of 212 bits each. The pulse stretching is performed independently by a transmitting node. The duration of the pulse varies according to a geometric distribution of probability $p = 0.5$. Thus, the pulse is larger than 1 slot time with a probability of 0.5, greater than 2 slots with a probability of 0.25, and so on. After the stretched pulse, nodes perform carrier sensing on an idle slot of 256 bits. The elimination phase ends with this idle slot (called elimination survival verification slot). Only contenders which simultaneously hold the highest access priority and select the longest stretched pulse survive the elimination phase and proceed to the yield phase. Here a random number of idle slots (0 to 9) is selected according to a geometric rate $r = 0.1$. The length of each slot is 168 bits. If a node detects no signal after a duration equal to the total number of idle slots has elapsed, the data packet is transmitted. Otherwise the node defers till the end of the current packet transmission. Note that the prioritization and contention intervals essentially comprise two listening periods separated by one transmission period (the access pattern), each of different lengths. The use of only one transmission period reduces radio switching overhead [5]. This is why the prioritization slots are based on decimal numbers and not binary digits (which will require less slots).

Synchronization is achieved by forcing each access pattern to start after the end of a packet transmission. If the medium is sensed idle for a minimum duration of 2,000 bits, EY-NPMA allows immediate access without the transmission of an access pattern. Like 802.11, unicast packets are individually acknowledged while broadcast/multicast packets are not acknowledged.

4.8.1 Intraforwarding

An interesting feature of the MAC protocol is that it permits relaying of packets among neighboring nodes using path discovery and packet forwarding algorithms, a process called intraforwarding. Intraforwarding allows messages sent to nodes which do not have direct radio connections to be forwarded by other nodes. This feature overcomes the limited range of independent wireless LANs and allows a wireless LAN to be extended without the need for access points or a wired backbone network. With the intraforwarding mechanism, adding new nodes to the network actually improves the reliability of a packet transmission since it can be routed to the destination node via more transmission paths. Under adverse propagation conditions, the intraforwarding algorithm allows a HIPER-LAN to be broken down into smaller subnetworks, each with a shorter range. Hence, an important advantage of the HIPERLAN Type 1 network is the ability to adapt to topology changes and to reroute packets when some wireless links fade. Note that it is impractical to enforce a unique identification with multiple independent wireless networks since coordination for this identification will be difficult.

The intraforwarding algorithm allows a relay node to forward a packet in two possible modes: point-to-point and broadcast transmission modes. To achieve this, each relay node maintains a routing table and a set of multipoint relays. The routing table indicates for each possible destination, the address of the closest relay node (next-hop relay node) to the destination and the path length. Hence, each packet carries two addresses: the final destination address and the next-hop relay address. Broadcast packets are forwarded only by multipoint relays. These relay nodes minimize the number of nodes forwarding the broadcast packets by taking advantage of the broadcast nature of radio transmission. Each broadcast packet in transit on a given relay node uses the same multipoint relay. A node which is not a multipoint relay does not forward a broadcast packet it receives from another node. The addresses of multipoint relays are not explicitly carried in the packet. Rather, the next-hop address field is replaced by the basic broadcast address. Multipoint relays are elected based on a minimum set of relay nodes that cover all remote nodes which are within two hops from the relay node (see Figure 4.28). This creates a two-level hierarchy on the network topology, hiding changes in distant parts of the network from local nodes.

To forward the packets, each node must be able to detect neighboring nodes with reliable and direct wireless links. Due to the unpredictable propagation characteristic of wireless transmission, all links must be checked both ways before they can be considered valid. The relay nodes employ a hop-by-hop algorithm which is invoked each time a packet is transmitted [6]. The algorithm offers the following advantages:

▶ The route used by a packet can be readily adapted to the latest topology while the packet is in transit;

▶ The headers of the packets are of fixed lengths and are much shorter compared to packets that employ source routing.

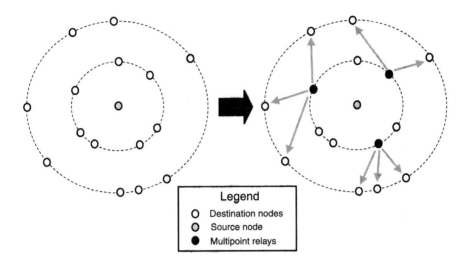

Figure 4.28 Election of multipoint relays.

Implementing intraforwarding requires fairly intensive computations to be performed by the mobile node. As mobile nodes have power, display, processing, and storage limitations, it may be preferable to offload much of such complex computational burden on the access point, which has fewer restrictions.

4.8.2 Hidden node

The MAC protocol overcomes the hidden node problem by making use of the fact that a receiving node can sense (detect) a signal even though it is unable to decode the signal (i.e., the sensing range can be much greater than the receiving range (see Figure 4.29)).

4.8.3 Quality of service

HIPERLAN offers "best effort" quality of service (QoS) with some control over the useful lifetime of a data packet. The standard defines a linear representation of priority comprising five access priority levels. Packet priority is a function of user-defined priority and residual (remaining) useful lifetime (see Table 4.12). While a lower priority packet is waiting, its residual lifetime will be decremented. The node may decide to increase the priority of a packet as its residual lifetime decreases. When the residual lifetime becomes zero and the packet has not been serviced, it will be discarded. Within the same priority class, first-come-first-served (FCFS)

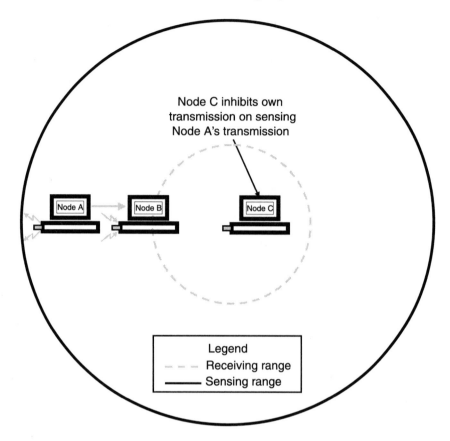

Figure 4.29   Hidden node elimination in the EY-NPMA protocol.

Table 4.12
Priority and Residual Lifetime

Normalized Residual Lifetime (NRL)	High User-Defined Priority	Low User-Defined Priority
NRL < 10 ms	0	1
10 ms ≤ NRL < 20 ms	1	2
20 ms ≤ NRL < 40 ms	2	3
40 ms ≤ NRL < 80 ms	3	4
NRL ≥ 80 ms	4	4

policy prevails. Hence, the MAC protocol in HIPERLAN Type 1 provides either best effort latency for isochronous traffic (e.g., voice, video) or best effort integrity for asynchronous traffic (e.g., data). Note that because a packet may be transmitted on more than 1 hop before it reaches the destination, the residual lifetime may be affected. Hence the priority mapping also takes into account the number of hops a packet has to travel.

4.8.4 Power management

The HIPERLAN defines two optional power conservation modes. The *p*-saver arranges the time it will be able to receive data while the *p*-supporter schedules the time it can transfer data to neighboring *p*-savers. These power-saving provisions are made through the low rate header.

4.8.5 Security

The HIPERLAN encryption-decryption scheme employs a single algorithm which requires an identical key and a matching initialization vector for both the encryption and decryption operations.

4.8.6 Commercial HIPERLAN Type 1 wireless LANs

During the development of the HIPERLAN standard, two projects, namely LAURA and HIPERION, have implemented various parts of the standard. Recently, the full capabilities of the HIPERLAN Type 1 technology were demonstrated by Thomson-CSF Detexis and STMicroelectronics at the inaugural HIPERLAN Alliance meeting held in May 1999. The systems were capable of supporting simultaneous high data-rate multimedia applications with quality of service provisions. Proxim have announced that their next generation high performance wireless LAN products will be based on HIPERLAN Type 1. Details can be found at http://www.proxim.com.

4.9 The WLIF OpenAir standard

A group of mobile computing product vendors formed the Wireless LAN Interoperability Forum (WLIF). Currently, the WLIF has 38 members.

Member companies deliver a wide range of interoperable wireless LAN products and services, thereby promoting the wireless LAN industry. The WLIF has published an OpenAir interface specification that enables independent parties to develop compatible products and has established a certification process for wireless LAN product interoperability. The WLIF specification is based on Proxim's 2.4 GHz FHSS wireless LAN introduced in early 1994. The system operates at a wireless data rate of 1.6 Mbps per hopping pattern. With 15 independent patterns available, aggregate data rates of up to 24 Mbps (15×1.6 Mbps) can be supported.

The OpenAir standard was completed by the WLIF in 1996. The WLIF is now actively working with the IEEE to establish interoperability between the OpenAir standard and the 802.11 standard, around which many members are developing products.

More information can be found at http://www.wlif.com.

4.10 The HomeRF SWAP standard

The high cost and inconvenience of adding new wires have inhibited the widespread adoption of home networking technologies. Recognizing this need, the HomeRF Working Group (HRFWG) was created in March 1998 to establish a set of wireless standards for interconnecting a broad range of electronic consumer products and personal computers anywhere in the home. The HRFWG, which includes 40 leading companies from the personal computer, consumer electronics, peripherals, communications, software, and semiconductor industries (e.g., Intel, Compaq, IBM, HP, Microsoft, Proxim, Motorola), has developed a specification for wireless communications in the home called the shared wireless access protocol (SWAP). Three subcommittees exist within the HRFWG. The HRFWG-Japan subcommittee was created to assist in defining the SWAP specification and ensure that it complies with local regulations. The group has also formed committees to plan future versions of SWAP that address wireless multimedia (SWAP-MM) and a lower cost alternative (SWAP-LITE).

HomeRF's shared wireless access protocol (SWAP) specification defines an over-the-air interface that is designed to support both wireless voice and data networking in and around the home. The standard can also interoperate with the public switched telephone network and the Internet. SWAP-compliant products operate in the 2.4 GHz frequency band using FHSS. The SWAP technology was derived from the existing digital enhanced cordless telephone (DECT) and IEEE 802.11 wireless LAN standards to enable a new class of home cordless services. As such, it provides voice support for DECT (adapted to 2.4 GHz) and TCP/IP support for 802.11. The 802.11 specifications have been relaxed to reduce product cost. For example, complex parts of the 802.11 MAC protocol such as PCF and RTS/CTS have been eliminated. SWAP supports TDMA (to provide delivery of interactive voice and other isochronous services) and CSMA/CA (for delivery of asynchronous high-speed packet data). The hybrid TDMA/CSMA frame structure is shown in Figure 4.30. A summary of the SWAP specifications is provided in Table 4.13.

4.10.1 Network configuration

The SWAP system can operate either as an ad hoc network or as a managed network under the control of a connection point. In an ad hoc network, where only data communication is supported, control of the network is distributed among all nodes. For time-critical communications

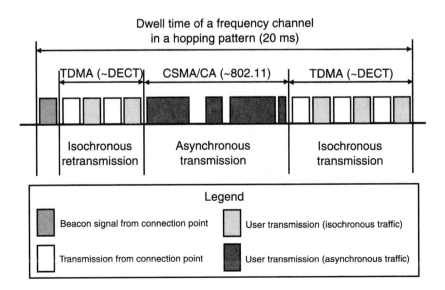

Figure 4.30 HomeRF packet structure.

Table 4.13
Main SWAP System Parameters

Parameter	Specification
Hopping rate	50 hop/s (same hopping patterns as 802.11)
Frequency range	2.4 GHz ISM band
Transmit radio power	100 min (20 dBm)
Data rate	1 Mbps (2-FSK), 2 Mbps (4-FSK)
Range	Up to 50 m
Number of nodes	Up to 127 devices per network
Voice connections	Up to 6 full duplex conversations, with error control
Data security	Blowfish encryption algorithm
Data compression	LZRW3-A algorithm
48-bit network identification	Enables operation of colocated networks

such as interactive voice, a connection point is required to coordinate the system. The connection point can be connected to a PC via a standard interface such as the universal serial bus (USB). The SWAP system also allows the connection point to support power management for extended battery life.

4.10.2 Applications

Some applications of the SWAP specification include:

▶ Accessing the Internet from anywhere in and around the home from portable display devices;

▶ Sharing files, modems, printers, and voice/Internet connections between PCs and other peripherals;

▶ Reviewing and forwarding voice, facsimile and e-mail messages;

▶ Activating home electronic systems by speaking a command to a PC-enhanced cordless handset.

More information is available at http://www.homerf.org.

4.11 The Bluetooth standard

The Bluetooth special interest group (SIG) was formed in February 1998 by major computer (Intel, IBM, Toshiba) and cellular telephone companies (Nokia, Ericsson) to provide wireless connectivity among mobile PCs, cellular phones, and other electronic devices. By early 1999, more than 700 companies had joined the Bluetooth SIG by signing the Bluetooth Adopters Agreement.

4.11.1 The need for the Bluetooth standard

Currently few mobile and portable devices communicate with each other. Those that do are mostly hardwired. The Bluetooth group has developed a standard that supports both voice and data traffic and provides access over different wide area networks. The standard defines one universal 1 Mbps full-duplex short-range radio link that connects up to eight portable digital devices. These devices include laptop computers, personal digital assistants, cellular phones, modems, wired telephones, printers, facsimile machines, headphones, digital cameras, and more. A typical application allows a portable computer to access a cellular phone (and thus the Internet) wirelessly. Connecting a headset to the cellular phone enables hands-free voice communications. When a digital camera is equipped with a Bluetooth radio module, images can be transferred immediately to a distant location via a mobile phone or a fixed telephone line. Besides allowing convenient sharing of information, Bluetooth may also generate innovative applications such as enabling a user to unlock his car using his mobile phone. First products conforming to the Bluetooth standard are expected in early 2000. Cost for these products are much lower than current wireless LAN components because they operate at low power.

4.11.2 Bluetooth specifications

All Bluetooth devices are peer entities that have identical implementations. However, when establishing a network one unit will act as a master and the others as slaves. A Bluetooth device transmits short packets in the

2.4 GHz ISM band using FHSS that changes frequency channels 1,600 times a second. The operating distance ranges from 10 cm to 10 m, but can be extended to more than 100 m by increasing the transmit power. Bluetooth can be built into virtually any mobile device by using a small-size (e.g., 1 cm × 3 cm) and low-cost radio module communication module. A typical Bluetooth implementation stack is shown in Figure 4.31. The functions of the various layers in the stack are described in Table 4.14.

4.11.3 Connection types

Bluetooth supports two types of connections:

1. Synchronous connection oriented (SCO);

2. Asynchronous connectionless (ACL).

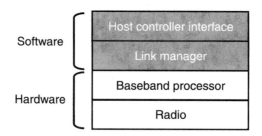

Figure 4.31 Bluetooth implementation stack.

Table 4.14
Functions of Stack Layers

Layer	Function
Radio	Short range microwave transceiver with clock and external transmitter.
Baseband processor	Specifies procedures needed to support the exchange of real-time voice and data as well as network information.
Link manager	Software function that executes protocols such as link establishment, network configuration, and authentication.
Host controller interface	Interface for Bluetooth host.

SCO packets are transmitted over reserved slots. Each packet is transmitted on a different frequency channel. At 1 Mbps and a hopping rate of 1,600 hop/s, this gives a maximum packet length of 625 bits or about 78 octets. A packet nominally covers a single slot, but can be extended to cover up to five slots. Once the connection is established, both master and slave units may send SCO packets without being polled. A SCO packet allows both voice and data transmission. However, only the data portion is retransmitted when corrupted.

The ACL connection supports both symmetric and asymmetric data traffic. The master unit controls the connection bandwidth and decides how much bandwidth is given to each slave. Slaves must be polled before they can transmit data.

Bluetooth can support one asynchronous data connection, up to three synchronous voice connections, or a connection which simultaneously supports a mixture of asynchronous data and synchronous voice. Each synchronous connection supports a data rate of 64 Kbps. An asynchronous connection supports 721 Kbps in the forward direction while permitting 57.6 Kbps in the reverse direction. Alternatively, it can support a symmetric connection of 432.6 Kbps.

4.11.4 Error correction

There are three error-correction schemes defined by Bluetooth:

1. 1/3 rate forward error correction (FEC) code;

2. 2/3 rate FEC code;

3. Automatic repeat request (ARQ) scheme for data.

The purpose of the FEC scheme is to cut down retransmissions. However, under reasonably error-free conditions, FEC creates an unnecessary overhead that reduces the throughput. Therefore, the packet definitions in Bluetooth have been kept flexible in order to accommodate the use of FEC. The packet header is always protected by a 1/3 rate FEC because it contains valuable link information. The ARQ scheme requires direct acknowledgment from the recipient.

4.11.5 Authentication and privacy

The Bluetooth standard provides authentication and encryption at the physical layer. Authentication is based on a challenge-response algorithm. Encryption is used to protect the privacy of a connection. Bluetooth uses a stream of ciphers for encryption.

4.11.6 Power consumption

Bluetooth devices consume very low power. Typical power consumptions are listed in Table 4.15.

4.11.7 Future developments for Bluetooth

The challenge for Bluetooth is how to make different networks with different speeds transparent to the end user. Applications need to be aware of the end-to-end speed of the wireless link because some communications devices (e.g., phone, two-way pager) do not have sufficient memory to buffer data. In addition, more than one application may be sharing the wireless link and a communicating device need not always be connected. The key to this problem is to build "connectivity aware" applications that are smart enough to detect the state of a current connection.

More information is available at http://www.bluetooth.com.

Table 4.15
Typical Bluetooth Power Consumption

Mode	Power
Standby	Less than 0.3 mA
Voice	8 to 30 mA
Data	5 mA (average)

4.12 The W3C and WAP standards

There are two associations currently standardizing a limited set of technical standards for mobile information access. The first is the mobile interest group within the World Wide Web Consortium (W3C) and the second is the Wireless Application Protocol (WAP) consortium.

The WAP Forum is dedicated to enabling advanced services and applications on wireless devices, such as mobile cellular phones. The W3C is dedicated to leading and advancing the development of the World Wide Web. Since June 1998, there is cooperation between the WAP Forum and the W3C in the area of mobile access to information on the Web. Instead of developing a diverging set of solutions, it is the intent of both groups to find common solutions.

In the area of Web technologies, the focus of the WAP Forum and the W3C overlap to a significant degree. Direct overlaps occur in the areas of intelligent proxies and protocol design, XML applications, and in content adaptation, e.g., through the use of vector graphics and style sheets). Future cooperation may also occur in the area of electronic payment, where the work of the two groups has potential overlap.

4.12.1 W3C

The W3C was founded in 1994 to develop an open forum for the evolution of Web technology. The consortium has more than 270 members from industry and academia. Like the Internet Engineering Task Force (IETF), the W3C specifications are based on sample code. In late 1997, it became evident that there was a considerable interest for access to the Web via mobile and wireless devices. As a result, a mobile access interest group was formed in April 1998 to investigate the impact of mobile access on the recommendations of the W3C.

More information is available at http://www.w3c.org.

4.12.2 WAP Forum

The WAP Forum, founded in December 1997, is an industry group comprising manufacturers, network operators, content providers, and application developers. It is dedicated to the goal of delivering content and applications to handheld wireless devices such as mobile phones, pagers, personal digital assistants, and other wireless terminals. Recognizing the value and utility of the World Wide Web, the WAP Forum has chosen to align certain components of its technology very tightly with the Internet and the Web.

Handheld wireless devices have special user interface constraints compared to personal computers. To enable a consistent application programming model, various levels of content scalability are required. However, current Web content is generally unsuitable for use on handheld wireless devices. Problems include:

▶ Input scalability where mobile devices feature a wide variety of input models, including numeric keypads with very few or no programmable softkeys;

▶ Output scalability where mobile devices have a broad range of visual display sizes, formatting, and other characteristics that differs from PC screens.

The WAP Forum has published a wireless application protocol (WAP) specification based on existing Internet standards such as XML and IP for wireless networks. The WAP specification aims to provide information and telephony services for digital mobile phones and other wireless terminals.

More information is available at http://www.wapforum.org.

4.13   The Infrared Data Association standard

The Infrared Data Association (IrDA) is an international organization of over 150 companies that creates and promotes interoperable, low-cost infrared data interconnection standards. Founded in 1993, IrDA produces standards that support a broad range of appliances as well as computing and communications devices. It has developed a short range, directed (line-of-sight) infrared standard suitable for personal computers, digital cameras, printers, handheld data collection devices, and so forth. The standard supports bidirectional infrared communications with up to eight peripherals simultaneously at a maximum data rate of 4 Mbps. The IrDA standard consists of a mandatory set of protocols and a set of optional protocols. The protocol stack is shown in Figure 4.32. The functions of these protocols are summarized in Tables 4.16 and 4.17.

More information is available at http://www.irda.org.

4.14 International wireless data associations

In this section, several international wireless data associations are introduced.

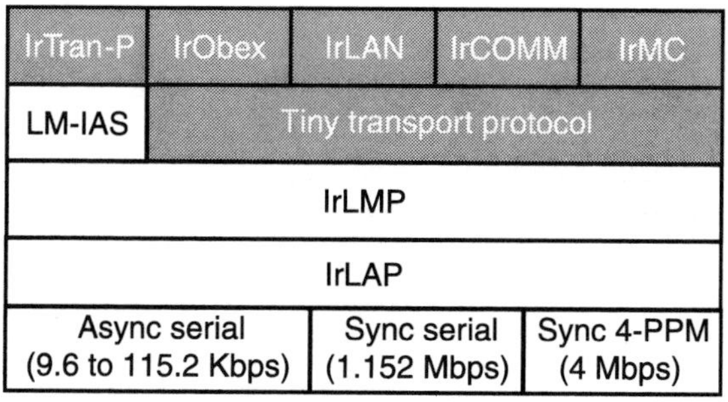

IrTran-P	IrObex	IrLAN	IrCOMM	IrMC
LM-IAS	Tiny transport protocol			
IrLMP				
IrLAP				
Async serial (9.6 to 115.2 Kbps)	Sync serial (1.152 Mbps)	Sync 4-PPM (4 Mbps)		

Figure 4.32 IrDA protocol stack.

Table 4.16
Mandatory IrDA Data Protocols

Protocol	Function
Physical Signaling Layer (PHY)	Provides data rates of 9600 bps to 4 Mbps at a range of 1 to 2 m. A low-power version relaxes the range to between 20 and 30 cm. Data packets are protected using CRC-16 for speeds up to 1.152 Mbps and CRC-32 at 4 Mbps.
Link Access Protocol (IrLAP)	Provides a device-to-device connection for the reliable, ordered delivery of data packets. It also covers device discovery procedures and handles hidden nodes.
Link Management Protocol (IrLMP)	Multiplexes multiple channels from an IrLAP connection. It also provides protocol and service discovery via the Information Access Service (IAS).

Table 4.17
Optional IrDA Data Protocols

Protocol	Function
Tiny Transport Protocol	Provides flow control on IrLMP connections with an optional segmentation/reassembly service.
IrCOMM	Provides serial/parallel port emulation for legacy COM applications, printing, and modem devices.
IrOBEX	Provides object exchange services similar to HTTP.
IrDA Lite	Provides methods of reducing the size of IrDA code while maintaining compatibility with full implementations.
IrTran-P	Provides image exchange using digital image capture devices/cameras.
IrMC	Specifies how mobile telephony and communication devices can exchange information e.g., phonebook, calendar, and message data. Also handles call control and real-time voice.
IrLAN	Describes a protocol used to support infrared access to LANs.

4.14.1   The Wireless LAN Alliance

The Wireless LAN Alliance (WLANA) is a nonprofit consortium of wireless LAN vendors established to provide information on specific applications, user experiences, current technologies, and future capabilities of wireless LANs. Current members of WLANA are listed in Table 4.18.

More information can be found at http://www.wlana.com.

4.14.2   The Wireless Data Forum

The Wireless Data Forum (WDF) is dedicated to promoting the benefits of wireless data with particular emphasis on networks based on the

Table 4.18
WLANA Members

3Com Corporation	Intersil Corporation	Norand Corporation
Advanced Micro Devices	IBM Corporation	Proxim
Andrew Corporation	Intermec Corporation	Raytheon Electronics
Aironet Wireless Communication	Lucent Technologies	Symbol Technologies
Cabletron	Nortel	Windata

Internet protocol (IP). WDF is composed of network service providers, wireless equipment vendors, computer software/hardware developers, and information services content providers.

More information is available at http://www.wirelessdata.org.

4.14.3 The Portable Computer and Communications Association

The Portable Computer and Communications Association (PCCA) was founded to provide a forum to enable disparate industries to meet and collaborate on issues related to the implementation of mobile computing.

More information is available at http://www.pcca.org.

4.15 Summary

The immense interest in wireless LANs has produced diverse technologies and standards. This is in stark contrast to other areas of communications marked by a convergence toward uniformity. Currently, the IEEE 802.11 standard is the only wireless LAN standard adopting a single MAC protocol that supports multiple physical layer specifications. The 802.11 MAC protocol also supports multiple data rates, mobility and power management, and solves problems due to the wireless medium such as high attenuation (CSMA/CA), openness (WEP), noise and interference (ACK), and the hidden node problem (RTS/CTS). The HIPERLAN Type 1 standard allows wireless LANs to be extended without the use of access points or a wired backbone network. This feature is a key advantage since the aim of a wireless network is to avoid cable. Medium access in HIPERLAN Type 1 is based on carrier-sensing and non pre-emptive priorities. A node automatically defers to the transmission of a higher priority packet. Emerging wireless industry standards for home and personal networking cater to a very specific set of applications, not the more general specifications envisioned by the IEEE 802.11 or HIPERLAN Type 1 standards.

References

[1]   Valadas, R., A. Tavares and A. Duarte, "The Infrared Physical Layer of the IEEE 802.11 Standard for Wireless Local Area Networks," *IEEE Communications Magazine*, Vol. 36, No. 12, December 1998, pp. 107–112.

[2]   Bing, B. and R. Subramanian, "A New Multiaccess Technique for Multimedia
       Wireless LANs," *Proceedings of the IEEE GLOBECOM*, Phoenix, AZ, USA,
       November 1997, pp. 1318–1322.

[3]   IEEE P802.11 *Information Technology—Telecommunications and Information
       Exchange Between Systems—Local and Metropolitan Area Networks—Specific
       Requirements, Part II: Wireless LAN Medium Access Control (MAC) and Physical
       Layer (PHY) Specifications,* November 1997.

[4]   EN 300 652, "Broadband Radio Access Networks (BRAN); High Performance
       Radio Local Area Network (HIPERLAN) Type 1; Functional Specification,"
       October 1998.

[5]   Jacquet, P. et al., "Priority and Collision Detection with Active Signaling—
       The Channel Access Mechanism of HIPERLAN," *Wireless Personal Communi-
       cations,* Vol. 4, No. 1, 1997, pp. 11–25.

[6]   Jacquet, P. et al., "Increasing Reliability in Cable-Free Radio LANs—Low
       Level Forwarding in HIPERLAN," *Wireless Personal Communication,* Vol. 4,
       No. 1, 1997, pp. 51–63.

Selected Bibliography

Halls, G., "HIPERLAN: The High Performance Radio Local Area Network Standard,"
Electronics and Communication Engineering Journal, December 1994, pp. 289–296.

LaMaire, R., A. Krishna, and P. Bhagwat, "Wireless LANs and Mobile Networking:
Standards and Future Directions," *IEEE Communications Magazine,* Vol. 34, No. 8,
August 1996, pp. 86–94.

Performance Evaluation of Wireless LANs

Ultimately, wired and wireless LANs will have to coexist since each has qualities that address particular user requirements. However, with cabled LANs such as Ethernet providing data rates of at least 10 Mbps, a key question that arises is whether wireless LANs are able to augment these higher speed LANs effectively. This chapter examines the characteristics of several commercial wireless LANs. The intent of the study is to characterize their behavior in terms of throughput and average delay under different degrees of network loading. It will be shown that the performance of wireless LANs under low to medium loads is comparable to traditional 10 Mbps Ethernet LANs. It will also be shown that quantitative results can be derived using a measurement technique based on network analyzers.

5.1 Performance evaluation techniques

Current methods of evaluating the performance of wireless LANs are largely based on analysis [1] and network simulation [2, 3]. While analytical studies and network simulation may provide valuable insights into the operation of wireless LANs, they cannot predict the actual performance of practical implementations with high accuracy. File transfer [4] is another method that can test the performance of a wireless LAN. However, measurements obtained from file transfer operations are limited by the need to specify the processor type, processor speed, and network operating system in order for meaningful comparisons to be made. The use of network analyzers for performance measurements is described in [5]. The utility of this technique is that it can determine the performance of a wireless LAN accurately due to the minimum overhead when data packets are sent and received at the lowest medium access control (MAC) sublayer.

5.2 Evaluation of IEEE 802.11 wireless LANs

Since the analyzers require a wired connection, wireless access points are required for the tests. In order to investigate the characteristics of different physical layers running at different data rates, the access points from Lucent Technologies (WavePOINT) [6] and Symbol Technologies (Spectrum24) [7] have been chosen. These products implement only the DCF MAC function and do not have the PCF function. The system specifications of the WavePOINT and Spectrum24 wireless LANs used in the performance evaluation are summarized in Table 5.1. More details can be found at http://www.wavelan.com and http://www.symbol.com. Currently, DFIR LANs conforming to the 802.11 standard are not available.

5.2.1 Theoretical calculations

For each unicast data packet that is successfully transmitted, additional overhead include:

Table 5.1
WavePOINT and Spectrum24 System Specifications

System Specification	WavePOINT	Spectrum24
Transmission Technology	DSSS	FHSS
Maximum Wireless Data Rate	2 Mbps	1 Mbps
Modulation	DQPSK	2-GFSK
Network Interface	MAC Sublayer	MAC Sublayer
Maximum Operating Range	60 to 430 m	50 to 600 m

▶ PLCP preamble/header;

▶ DCF interframe space (DIFS);

▶ Short interframe space (SIFS);

▶ ACK packet;

▶ RTS and CTS packets (if virtual carrier sensing is activated).

The overhead for the transmission of data, ACK, RTS, and CTS packets is shown in Figure 5.1. For broadcast and multicast packets, overhead associated with the SIFS and the ACK, RTS, and CTS packets is not applicable. Note that the data, ACK, RTS, and CTS packets all require the PLCP preamble and header, which is a fixed overhead.

The interframe spaces (or time gaps), together with the overhead for the PLCP preamble/header and the ACK, RTS, and CTS packets are summarized in Table 5.2. Although the list in Table 5.2 constitutes the main overhead, it represents the minimum required for the transmission of one data packet. Extra overhead due to the transceiver transmission delays (e.g., ramp-on and ramp-off times) and other implementation-dependent delays has not been included. For FHSS wireless LANs, the time to hop between frequency channels is specified at 224 μs. This hopping time is ignored because a packet is very likely to be transmitted within the large dwell time of a single frequency channel (see Section 2.3). Retransmission due to collisions is also not considered since the measurements focus on the performance of a single wireless link between two access points. Note the substantial overhead when the RTS/CTS feature is activated.

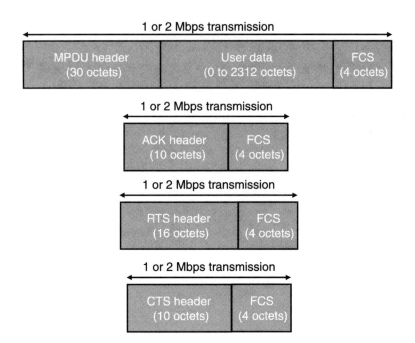

Figure 5.1   Overhead for data, ACK, RTS, and CTS packets.

Table 5.2
Overhead for the Transmission of One Data Packet

Overhead for Wireless Transmission	WavePOINT	Spectrum24
DIFS	50 μs	128 μs
SIFS	10 μs	28 μs
PLCP Preamble/Header	24×8 bits/1 Mbps = 192 μs	16×8 bits/1 Mbps = 128 μs
MPDU Header/FCS	34×8 bits/2 Mbps = 136 μs	34×8 bits/1 Mbps = 272 μs
ACK Header/FCS	14×8 bits/2 Mbps = 56 μs	14×8 bits/1 Mbps = 112 μs
RTS Header/FCS	20×8 bits/2 Mbps = 80 μs	20×8 bits/1 Mbps = 160 μs
CTS Header/FCS	14×8 bits/2 Mbps = 56 μs	14×8 bits/1 Mbps = 112 μs
Total Overhead (Broadcast)	378 μs	528 μs
Total Overhead (Unicast with no RTS/CTS)	636 μs	796 μs
Total Overhead (Unicast with RTS/CTS)	1176 μs	1380 μs

Consider the transmission of a maximum-length Ethernet packet (unicast with no RTS/CTS) through the WavePOINT and Spectrum24 access points. Although the maximum length of an Ethernet packet is actually 1,526 octets, overhead due to the preamble (7 octets) and the start frame delimiter, or SFD (1 octet), is excluded from the following calculations which deal exclusively with wireless transmission. This is because the preamble and SFD of an Ethernet packet do not serve any useful purpose for wireless transmission (recall that the wireless PLCP packet has its own preamble and header). Hence, they are typically removed or compressed at the transmitting end to reduce wireless transmission overhead (see for example [8]) and then reinserted back at the receiving end.

The time to transmit one maximum-length Ethernet packet of 1,518 octets using WavePOINT is

$$\frac{1518 \times 8 \text{ bits}}{2 \text{ Mbps}} = 6{,}072 \text{ } \mu s$$

and the wireless throughput becomes

$$\frac{1518 \times 8 \text{ bits}}{6072 \text{ } \mu s + 636 \text{ } \mu s} = 1.810 \text{ Mbps}$$

For Spectrum24, the transmission time for one maximum-length Ethernet packet is

$$\frac{1518 \times 8 \text{ bits}}{1 \text{ Mbps}} = 12{,}144 \text{ } \mu s$$

and the wireless throughput becomes

$$\frac{1518 \times 8 \text{ bits}}{12144 \text{ } \mu s + 764 \text{ } \mu s} = 0.941 \text{ Mbps}$$

It should be pointed out that although Spectrum24 incurs more overhead (764 μs) for the transmission of one unicast data packet compared to WavePOINT (636 μs), the wireless throughput efficiency (94.1%) is actually higher than that of WavePOINT (90.5%). This is attributed to the fact that the PLCP preamble and header have to be transmitted at 1 Mbps (even though the MAC portion may be transmitted at a data rate higher than 1 Mbps, as in the case for WavePOINT).

Hence, the overhead for the transmission of the PLCP preamble and header of a data packet using WavePOINT appears large relative to the transmission time for the MAC portion [9]. Since the overhead is constant for every data packet transmitted, the difference in throughput efficiencies becomes more pronounced with shorter data packets (see Tables 5.3 and 5.4).

Table 5.3
Theoretical Wireless Throughput Efficiency for WavePOINT (2 Mbps)

Ethernet Packet Length (octets)	Ethernet Packet Transmission Time (μs)	Broadcast Throughput Efficiency	Unicast Throughput Efficiency with no RTS/CTS	Unicast Throughput Efficiency with RTS/CTS
1518	6,072	94.1%	90.5%	83.8%
1504	6,016	94.1%	90.4%	83.6%
1004	4,016	91.4%	86.3%	77.3%
504	2,016	84.2%	76.0%	63.2%
204	816	68.3%	56.2%	41.0%
104	416	52.4%	39.5%	26.1%
64	256	40.4%	28.7%	17.9%

Table 5.4
Theoretical Wireless Throughput Efficiency for Spectrum24 (1 Mbps)

Ethernet Packet Length (octets)	Ethernet Packet Transmission Time (μs)	Broadcast Throughput Efficiency	Unicast Throughput Efficiency with no RTS/CTS	Unicast Throughput Efficiency with RTS/CTS
1518	12,144	95.8%	93.8%	89.8%
1504	12,032	95.8%	93.8%	89.7%
1004	8,032	93.8%	91.0%	85.3%
504	4,032	88.4%	83.5%	74.5%
204	1,632	75.6%	67.2%	54.2%
104	832	61.2%	51.1%	37.6%
64	512	49.2%	39.1%	27.1%

5.2.2 Measurement setups

The performance tests rely on a network analyzer to generate Ethernet packets continuously so that the throughput and average delay of the wireless LANs can be measured. The analyzer enables the length of an Ethernet packet and the interframe space (IFS) or the interarrival time between consecutive packets to be varied accurately at the MAC sublayer. The IFS includes the standard Ethernet IFS of 9.6 μs. Thus, the network traffic load may be changed to any desired level.

The measurement setups are shown in Figure 5.2. The wired measurement facilitates comparison with the wireless measurement. In the wireless measurement, the analyzers are connected to the access points using short wired segments. One of the analyzers is configured to be a transmitter while the other operates in the receiving mode. It is expected that the wireless transceivers of the access points will not be able to cope with the wired Ethernet transmission from the transmitting analyzer (which operates at a maximum data rate of close to 10 Mbps without wireless overhead [10, 11]) under heavy network loading. Hence, the maximum wireless throughput may be obtained when one or more missing Ethernet packets are detected by the receiving analyzer. For example, if 100 Ethernet packets have been transmitted by analyzer A but only 96 packets are received by analyzer B, it implies that 4 Ethernet packets have been discarded by the wireless transceivers because they are unable to handle the transmission rate from analyzer A. A network

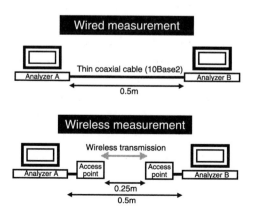

Figure 5.2 Experimental configurations for wired and wireless measurements.

analyzer is passive when operated in the receiving mode. As such, error recovery is not performed on corrupted packets. The analyzer also disables its receiver while generating traffic.

The short distance between the wireless access points minimizes the signal propagation delay and the impact of impairments related to the wireless medium (e.g., multipath propagation). This in turn reduces the number of parameters needed for analysis and provides a more accurate investigation into the performance limits of the access points. Although additional delays (e.g., delays due to carrier sensing, transmission of preamble/SFD) may be incurred in the wired segments, these delays are minimal since the segments are short and packets are transmitted at the Ethernet data rate of about 10 Mbps (which is 5 to 10 times higher than the data rate of the wireless transceiver). Note that contention does not occur on the wired segments since only one analyzer is attached to each segment. Thus, random delays due to contention in the wired segments are avoided.

Before the actual measurements are carried out, Ethernet packets with different data patterns are transmitted and analyzed. The average delay and the wireless throughput of the access points have been found to be independent of the data pattern.

5.2.3   Measured results

In the graphical plots and tabulated results presented in this section, three parameters (i.e., the average delay, the maximum wireless throughput and the minimum loss-free IFS) are defined as follows. The average delay represents the average time a new Ethernet packet arrives at the transmitting analyzer to the time the complete packet is successfully received by the receiving analyzer; it excludes the IFS. The maximum wireless throughput is computed using the ratio of the length of a transmitted Ethernet packet to the minimum average time to receive the packet. The minimum average time is calculated by measuring the minimum time to receive a series of packets with no packet loss detected and then averaging it over all the packets transmitted. The minimum loss-free IFS corresponds to the minimum IFS when no packet loss is detected by the receiving analyzer.

Figures 5.3, 5.4, and 5.5 show that both WavePOINT and Spectrum24 exhibit high latency under heavy traffic load. This can be attributed to

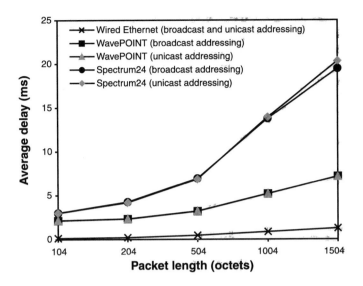

Figure 5.3 Average delay performance (IFS = 0.1 ms).

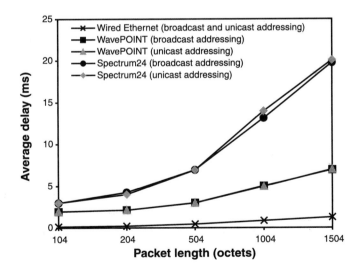

Figure 5.4 Average delay performance (IFS = 0.3 ms).

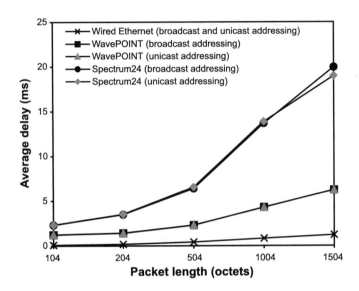

Figure 5.5   Average delay performance (IFS = 1 ms).

the need for the wireless transceivers to buffer arriving packets first before actual transmission so as to keep up with the high load. In Figure 5.6, the average delay performance of WavePOINT becomes identical to that of wired Ethernet. In Figure 5.7, both WavePOINT and Spectrum24 have similar average delay performances as wired Ethernet. These five figures indicate that as the traffic load decreases, the average delay performance of the wireless LANs can approach that of wired Ethernet. Figure 5.8 shows that WavePOINT and Spectrum24 are capable of maximum wireless throughputs of about 1.67 Mbps and 0.58 Mbps respectively. The average delay and throughput values are practically unchanged regardless of whether packets are sent with broadcast or unicast addresses. The wireless throughput performance of WavePOINT degrade considerably when short Ethernet packets are transmitted. This can be attributed to the overhead of the PLCP preamble and header, as explained in Section 5.3.

An interesting characteristic of the wireless throughput for Spectrum24 is that the maximum throughput is reached when Ethernet packets of 504 octets are sent. Unlike WavePOINT, increasing packet lengths do not improve the throughput further. This characteristic may be caused by the fragmentation of long packets (e.g., 1,004 and 1,504

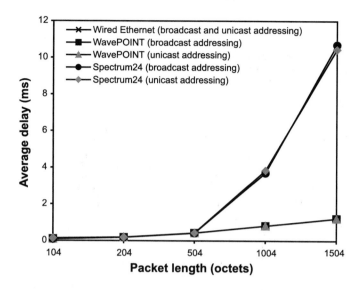

Figure 5.6 Average delay performance (IFS = 10 ms).

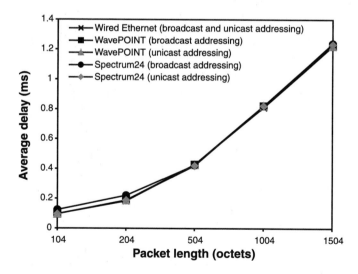

Figure 5.7 Average delay performance (IFS = 100 ms).

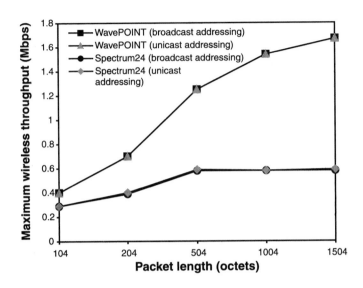

Figure 5.8 Maximum wireless throughput performance.

octets) into packets of 504 octets. The Spectrum24 access point operates with a fixed fragmentation threshold of 504 octets. The additional overhead due to fragmentation seemed to offset the gain in wireless transmission reliability. Lucent Technologies has indicated that the fragmentation feature will be added in future firmware upgrades of their wireless LANs.

Tables 5.5 and 5.6 show samples of the measured delays for WavePOINT and Spectrum24. In these measurements, Ethernet packets are transmitted with broadcast addresses. The time to transmit 100 packets from the analyzer is calculated theoretically, taking into account the length of the Ethernet packet (preamble and SFD inclusive) and the IFS. By comparing the receiving delays with the transmission times, the receiving delays indicated in bold suggest that packet buffering at the access points may be responsible for the relatively constant delays, even as the IFS decreases. To verify this, the number of packets transmitted varied from 100 to 2,000, the minimum loss-free IFS measured and the corresponding throughput for WavePOINT calculated in Table 5.7. Clearly, the maximum wireless throughput remains unaffected by changes in the minimum loss-free IFS. Thus, the effect of packet buffering at the wireless access point is to increase but smooth out the delay when the transmission rate from the wired network is high. The maximum

Table 5.5

Delay Measurements for WavePOINT (Broadcast Addressing)

Ethernet Packet Length (octets)	Interframe Space (ms)	Time to Transmit 100 Ethernet Packets from Analyzer A (s)	Time to Receive 100 Ethernet Packets on Analyzer B (s)
1504	0.1	0.13096	**0.7264**
1504	1.0	0.22096	**0.7230**
1504	10.0	1.12096	1.1220
1504	100.0	10.12096	10.1223
1004	0.1	0.09096	**0.5265**
1004	1.0	0.18096	**0.5306**
1004	10.0	1.08096	1.0816
1004	100.0	10.08096	10.0821
504	0.1	0.05096	**0.3290**
504	1.0	0.14096	**0.3299**
504	10.0	1.04096	1.0425
504	100.0	10.04096	10.0428
204	0.1	0.02696	**0.2417**
204	1.0	0.11696	**0.2418**
204	10.0	1.01696	1.0185
204	100.0	10.01696	10.0188
104	0.1	0.01896	**0.2238**
104	1.0	0.10896	**0.2238**
104	10.0	1.00896	1.0096
104	100.0	10.00896	10.0096

wireless throughput is reached when the buffer space becomes depleted and further packet arrivals are discarded. Note that besides acting as an interface between a high-speed wired network and a lower-speed wireless LAN, packet buffering also serves other functions. For instance, it is normally employed by wireless access points to support roaming (when mobile nodes move among wireless coverage areas) as well as to provide temporary storage of packets addressed to nodes operating in the power saving mode. Packet buffering is particularly useful in maintaining an ongoing network connection when a mobile node temporarily moves out of the coverage area of the access point.

Table 5.6
Delay Measurements for Spectrum24 (Broadcast Addressing)

Ethernet Packet Length (octets)	Interframe Space (ms)	Time to Transmit 100 Ethernet Packets from Analyzer A (s)	Time to Receive 100 Ethernet Packets on Analyzer B (s)
1504	0.1	0.13096	**1.9574**
1504	1.0	0.22096	**2.0944**
1504	10.0	1.12096	**2.0702**
1504	100.0	10.12096	10.1239
1004	0.1	0.09096	**1.3863**
1004	1.0	0.18096	**1.3365**
1004	10.0	1.08096	**1.3713**
1004	100.0	10.08096	10.0821
504	0.1	0.05096	**0.7044**
504	1.0	0.14096	**0.7280**
504	10.0	1.04096	1.0414
504	100.0	10.04096	10.0424
204	0.1	0.02696	**0.4363**
204	1.0	0.11696	**0.4490**
204	10.0	1.01696	1.0185
204	100.0	10.01696	10.0220
104	0.1	0.01896	**0.3098**
104	1.0	0.10896	**0.3312**
104	10.0	1.00896	1.0094
104	100.0	10.00896	10.0124

The minimum loss-free IFS characteristics (broadcast addressing) for both access points are plotted in Figures 5.9 and 5.10. It can be observed that as the number of transmitted packets increases, the minimum loss-free IFS increases. This phenomenon is expected since the buffer space runs out faster when more packets are transmitted. A closer examination reveals a significant increase in the minimum loss-free IFS when packets longer than 504 octets are transmitted. This suggests that the wireless packet transmission rates of both access points are optimized to operate with short packets (of lengths that lie roughly below 504 octets). Long packets require substantial buffering space at the access

Table 5.7
Throughput Measurements for WavePOINT (Broadcast Addressing)

Ethernet Packet Length (octets)	Number of Ethernet Packets Transmitted from Analyzer A	Minimum Loss-Free Interframe Space (ms)	Minimum Time to Receive All Transmitted Packets on Analyzer B (s)	Maximum Wireless Throughput (Mbps)
1504	100	0.04	0.7237	1.67
1504	500	3.66	3.6210	1.67
1504	1000	4.85	7.2590	1.67
1504	1500	5.25	10.8788	1.67
1504	2000	5.45	14.4900	1.67
1004	100	0.04	0.5244	1.54
1004	500	2.07	2.6264	1.54
1004	1000	3.24	5.2555	1.54
1004	1500	3.65	7.8760	1.54
1004	2000	3.83	10.4937	1.54
504	100	0.04	0.3266	1.25
504	500	0.39	1.6403	1.25
504	1000	1.61	3.2631	1.26
504	1500	2.01	4.8914	1.26
504	2000	2.22	6.5276	1.25
204	100	0.04	0.2408	0.70
204	500	0.23	1.2073	0.70
204	1000	1.23	2.4194	0.70
204	1500	1.57	3.6357	0.70
204	2000	1.74	4.8488	0.70
104	100	0.04	0.2245	0.40
104	500	0.05	1.1162	0.40
104	1000	1.14	2.2372	0.40
104	1500	1.47	3.3650	0.40
104	2000	1.64	4.4847	0.40

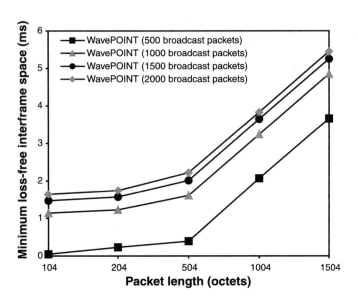

Figure 5.9   Minimum loss-free IFS (WavePOINT).

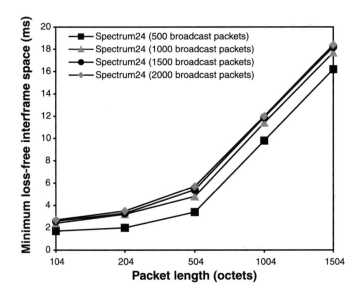

Figure 5.10   Minimum loss-free IFS (Spectrum24).

points for a fixed buffer space, the wireless packet transmission rate becomes restricted considerably.

The effect of RTS/CTS on the maximum wireless throughput for WavePOINT is shown in Table 5.8. The RTS/CTS packets are generated by the wireless transceivers (and not by network analyzers). By comparing the values in Tables 5.7 and 5.8, it is evident that the transmission of RTS/CTS packets has little impact on the actual performance of an 802.11 wireless LAN. Once again, this may be due to packet buffering, which helps to regulate the wireless throughput and average delay under heavy network loading. Note that the throughput efficiencies for packet lengths 504, 1,004, and 1,504 octets are very close to that computed in Table 5.3. For shorter packet lengths (i.e., 104 and 204 octets), the slight discrepancies may be caused by the additional overhead that has been omitted in the calculations for Table 5.3 becoming more significant for these packets.

5.3 The non-802.11 WavePOINT wireless LAN

In order to compare the performance of the standard and nonstandard wireless LANs, a performance evaluation of the 2.4 GHz non-802.11 WavePOINT has been carried out. The 2 Mbps DSSS access point operates with the same 11-chip Barker code as specified in the 802.11 standard. It also implements a variant of nonpersistent CSMA/CA which is very similar to the MAC protocol adopted by the 802.11 standard.

Table 5.9 shows the wireless throughput performance provided by the manufacturer. It can be observed that the wireless throughput decreases as the packet length decreases.

The packet format is shown in Figure 5.11. The network identification is a 4-digit hexadecimal code that logically connects wireless devices belonging to the same wireless coverage. Without the network identification, a third party can neither connect to the wireless network nor capture a message.

The efficiency for transmitting one maximum-length Ethernet packet (1,518 octets) using the 2 Mbps non-802.11 WavePOINT is

$$\frac{1,518}{1,518+37.5} \times 100\% = 97.6\%$$

Table 5.8
Throughput Measurements for WavePOINT
(Unicast Addressing with RTS/CTS)

Ethernet Packet Length (octets)	Number of Ethernet Packets Transmitted from Analyzer A	Minimum Loss-Free Interframe Space (ms)	Minimum Time to Receive All Transmitted Packets on Analyzer B (s)	Maximum Wireless Throughput (Mbps)
1504	100	0.04	0.7263	1.67
1504	500	3.67	3.6242	1.67
1504	1000	4.85	7.2551	1.67
1504	1500	5.25	10.8680	1.67
1504	2000	5.45	14.5006	1.67
1004	100	0.04	0.5276	1.53
1004	500	2.07	2.6301	1.54
1004	1000	3.25	5.2507	1.54
1004	1500	3.65	7.8696	1.54
1004	2000	3.84	10.4999	1.54
504	100	0.04	0.3340	1.23
504	500	0.39	1.6461	1.24
504	1000	1.61	3.2677	1.25
504	1500	2.01	4.8985	1.25
504	2000	2.22	6.5343	1.25
204	100	0.04	0.2422	0.70
204	500	0.23	1.2096	0.70
204	1000	1.24	2.4233	0.70
204	1500	1.57	3.6450	0.70
204	2000	1.74	4.8624	0.70
104	100	0.04	0.2246	0.40
104	500	0.06	1.1186	0.40
104	1000	1.14	2.2438	0.40
104	1500	1.47	3.3695	0.40
104	2000	1.64	4.4941	0.40

Table 5.9
WavePOINT Wireless Throughput Performance [12]

Ethernet Packet Length (octets)	Forwarding Rate (packet/s)	Wireless Throughput (Mbps)
64	2,050	1.05
512	390	1.60
1518	140	1.70

Note: Filtering rate 6,000 packet/s at 64 octets per packet.

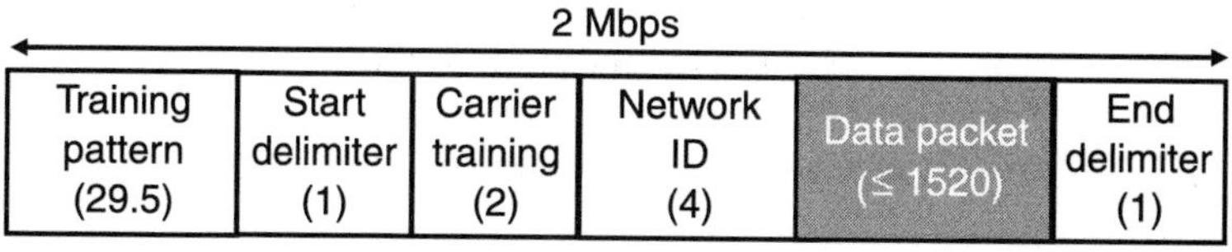

2 Mbps					
Training pattern (29.5)	Start delimiter (1)	Carrier training (2)	Network ID (4)	Data packet (≤ 1520)	End delimiter (1)

Figure 5.11 Packet format for the non-802.11 WavePOINT.

For the 2 Mbps 802.11 WavePOINT, the maximum efficiency is

$$\frac{1{,}518}{1{,}518+48} \times 100\% = 96.9\%$$

5.3.1 Measured results

As shown in Figures 5.12, 5.13 and 5.14, broadcast packet addressing can degrade the performance of the non-802.11 WavePOINT. This is in contrast to the performance of the 802.11 WavePOINT. The access point achieves a wireless throughput of about 0.8 Mbps (broadcast addressing) and 1.2 Mbps (unicast addressing) when Ethernet packets of 104 octets are transmitted (see Figure 5.15). These throughput values are 2 to 3 times higher than the 802.11 WavePOINT. The maximum wireless throughput for both WavePOINTs, however, are roughly the same. The results suggest that the 802.11 WavePOINT performs much better when longer packets (e.g., packet lengths above 504 octets) are transmitted.

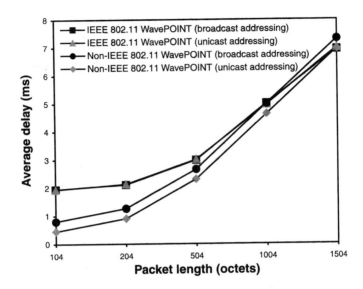

Figure 5.12   Average delay comparison (IFS = 0.3 ms).

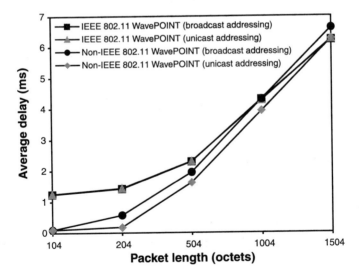

Figure 5.13   Average delay comparison (IFS = 1 ms).

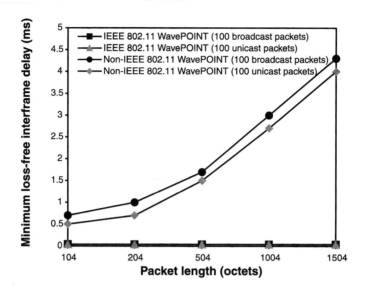

Figure 5.14 Minimum loss-free IFS comparison.

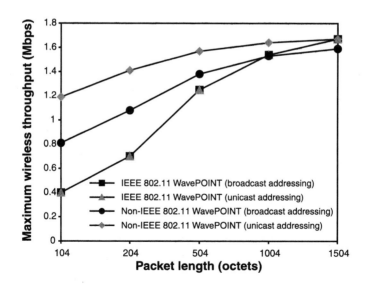

Figure 5.15 Maximum wireless throughput comparison.

5.4 The Motorola Altair Plus wireless LAN

The previous sections have examined the performance of spread-spectrum wireless LANs. In this section, a wireless LAN that employs narrowband radio transmission is evaluated. The Motorola Altair Plus wireless LAN operates in the licensed 18 to 19 GHz (18.82 to 18.87 GHz and 19.16 to 19.21 GHz) frequency band. This band is segmented into ten 10 MHz channels, each providing 15 Mbps of transmission capacity. The narrowband wireless LAN system exploits the limited range of radio transmission to reuse the same frequency channels spatially, much like in cellular networks. Hence, the network capacity in large coverage areas can be larger than the 15 Mbps data rate. However, this requires frequency planning which divides the network into distinct coverage areas. To minimize interference, nonoverlapping frequency channels must be assigned to adjacent coverage areas.

The Altair wireless LAN incorporates many advanced implementation concepts that may well prove to be indispensable in broadband wireless networks. Although the Altair system is primarily designed for Ethernet data traffic, it implements a reservation-based multiple access technique that can allow different types of traffic to be accommodated (e.g., a combination of voice, video, and data traffic). Besides using gallium arsenide technology, Altair also employs six-sector directional antennas in each radio transceiver. Radio energy radiates only in a particular direction, providing gain along the intended direction and attenuation in the undesired directions. This helps to increase system capacity while reducing interference and transmit power requirements. Selection algorithms at the antennas assess the received signal strength and quality from 36 different paths every 24 ms (see Figure 5.16). This gives the system a high probability of finding a path that is not corrupted by fading or interference, thereby mitigating undesirable multipath effects.

Altair operates with a centralized base unit called the control module (CM) that relays packets to and from one or more user modules (UMs) in the network, effectively removing the hidden-node problem. The CM also acts as a central controller that regulates communications. UMs contend for short request slots (using slotted ALOHA) in order to reserve larger, contention-free data slots. Altair uses a 4-level continuous phase frequency shift keying (4-CPFSK) modulation scheme with time division

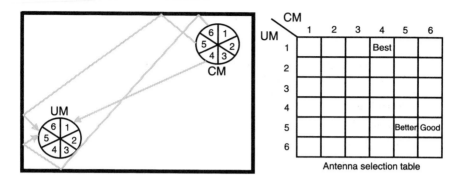

Figure 5.16 Antenna selection in Altair.

duplexing (TDD). The TDD scheme facilitates transmission of control information and user data between the CM and UM. More information on Altair can be found in [13, 14].

Altair operates with an Ethernet MAC interface. It is based on wire replacement concepts for desktop PCs rather than for portables. The packet format is shown in Figure 5.17. Clearly, packet fragmentation is adopted in the Altair wireless LAN which restricts each transmitted Ethernet data packet to a maximum length of 380 octets. Note that the maximum throughput efficiency for transmitting one maximum length Ethernet packet is $1,518/4,031.25 \times 100\% = 37.66\%$. For a wireless data rate of 15 Mbps, this efficiency is equivalent to 5.65 Mbps.

From Table 5.10, several observations can be made. It is clear that Altair's wireless throughput decreases as packet length decreases. Unlike WavePOINT, the packet forwarding rate is constant for a range of packet lengths. The packet handling capabilities increases as broadcast traffic decreases. In the most severe case, a broadcast packet may be transmitted sequentially from all six antenna sectors, resulting in a transit time of 12 ms [15]. Note that the values provided can only be used as a general guide

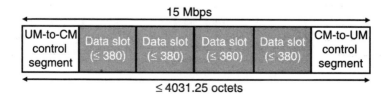

Figure 5.17 Packet format for the Altair wireless LAN.

Table 5.10
Altair Wireless Throughput Performance [14]

Ethernet Packet Length (octets)	Forwarding Rate (packet/s)	Wireless Throughput (Mbps)
64 to 380	850	0.41 to 2.58
381 to 760	705	2.15 to 4.29
761 to 1140	580	3.53 to 5.29
1141 to 1518	465	4.24 to 5.65

since they do not account for possible collisions (and retransmission) or the addressing of a packet (which can be broadcast or unicast). Furthermore, no details about the network setup (e.g., direction of transmission and distance) are provided.

5.4.1 Measurement setups

The point-to-point wireless performance measurement setup is similar to Figure 5.2 (Section 5.3). However, for Altair, one of the access points must be a CM. In the point-to-multipoint wireless performance measurement analyzers A and B are linked up by two UMs. Another analyzer, C, is connected to the CM (see Figure 5.18). Ethernet packets with varying interframe delay and lengths are transmitted simultaneously from the UMs to the CM.

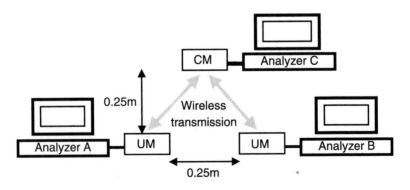

Figure 5.18   Point-to-multipoint measurement for the Altair wireless LAN.

5.4.2 Point-to-point measured results

Like WavePOINT, Figure 5.19 shows that Altair also exhibits high latency under heavy traffic load. When a single UM intends to communicate with a CM, the need to reserve for transmission time adds to this latency. When two UMs wish to communicate with a CM, the need to contend before reserving transmission time further increases the average delay. It is also evident that broadcast addressing severely degrades the average delay although this effect does not seem to apply for the case when the transmission direction is from CM to UM. In Figure 5.20, the average delay is identical for both wired and wireless transmission when a CM transmits to one UM. In Figures 5.21 and 5.22, the average delay for wired transmission matches that of wireless transmission when a single UM communicates.

Results from Figure 5.23 indicate that the highest wireless throughput of about 5.62 Mbps is attained when a CM transmits directly to one UM using 1,504-octet packets. This throughput is virtually independent of the type of addressing employed. It can be concluded that the CM controls the network of UMs, and does not have to reserve for time slots during transmission. When the direction of transmission is reversed (i.e., from

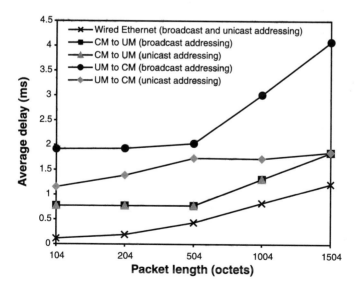

Figure 5.19 Average delay performance (IFS = 0.3 ms).

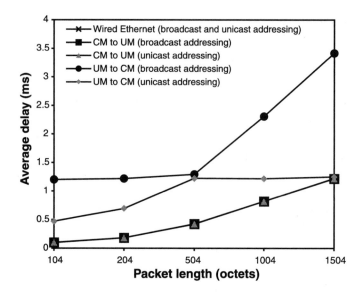

Figure 5.20   Average delay performance (IFS = 1 ms).

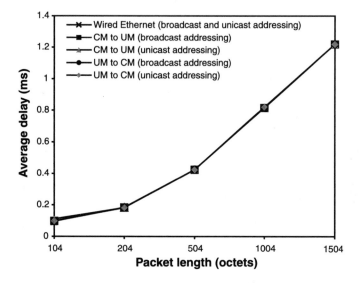

Figure 5.21   Average delay performance (IFS = 10 ms).

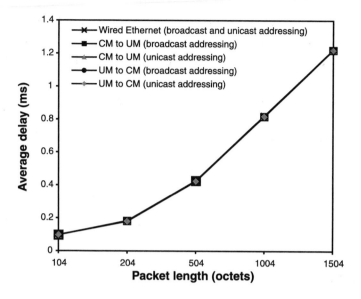

Figure 5.22   Average delay performance (IFS = 100 ms).

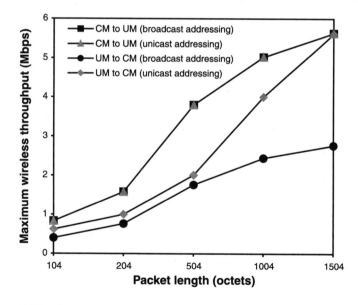

Figure 5.23   Maximum wireless throughput performance.

UM to CM) and broadcast addressing is used, the throughput is significantly lowered. This drastic drop in throughput can be attributed to the fact that the UM needs to reserve for time slots before transmitting broadcast information. It can be observed from Figure 5.24 that the minimum loss-free IFS for Altair does not change significantly.

5.4.3 Point-to-multipoint measured results

Figures 5.25 to 5.30 illustrate the performance of the Altair system when contention occurs between 2 UMs. Clearly, collisions due to contention can degrade the average delay and wireless throughput significantly.

5.5   Summary

The practical network performances of several commercially-developed wireless LANs have been measured and analyzed. A number of tests conducted on the wireless LANs yielded important characteristics such as throughput and average delay under various network loads. In most cases, the wireless LAN performance characteristic is plotted together

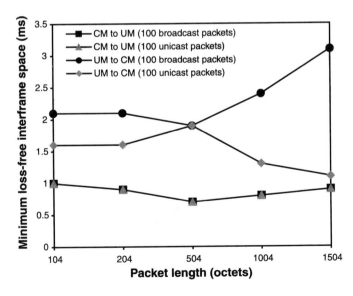

Figure 5.24 Minimum loss-free IFS comparison.

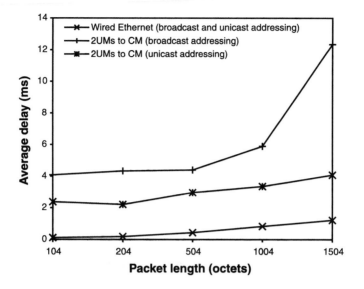

Figure 5.25 Average delay performance (IFS = 0.3 ms).

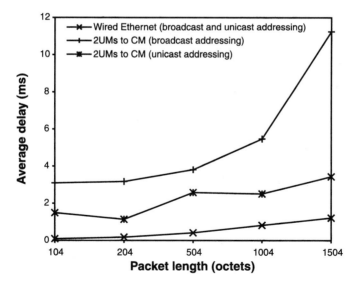

Figure 5.26 Average delay performance (IFS = 1 ms).

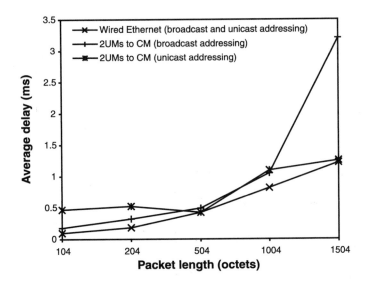

Figure 5.27   Average delay performance (IFS = 10 ms).

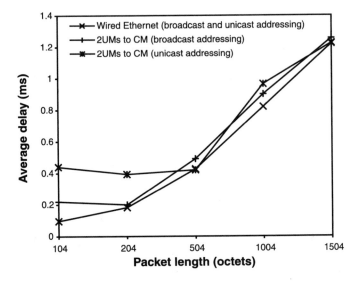

Figure 5.28   Average delay performance (IFS = 100 ms).

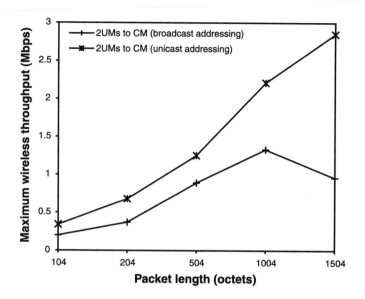

Figure 5.29 Maximum wireless throughput performance.

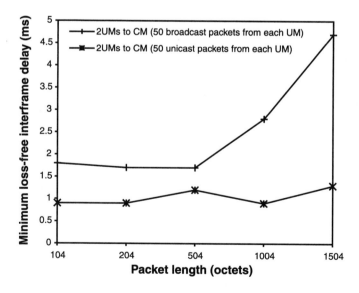

Figure 5.30 Minimum loss-free IFS comparison.

with that of wired Ethernet, thus facilitating wired and wireless perform-ance comparison.

For 802.11 compliant LANs, the results suggest that packet buffering and packet fragmentation are crucial factors in determining the perfor-mance of a wireless LAN. While the length of a data packet and the wireless data rate also affect a wireless LAN's transmission capabilities, the performance of an 802.11 wireless LAN is generally unaffected by the type of packet addressing and the use of reservation packets such as RTS and CTS. The average delay performance for both systems is similar to that of wired Ethernet for an interframe space in the region of 10 ms and above (for WavePOINT) and 100 ms and above (for Spectrum24).

For non-802.11 compliant LANs, the type of addressing and the length of an Ethernet packet are important factors in determining a wireless LAN's transmission capabilities. Specifically, the throughput of a wireless LAN increases as the packet length increases and as the amount of broadcast traffic decreases. For Altair, the direction of transmission (i.e., from CM to UM or from UM to CM) also has an enormous impact on the throughput and average delay. The average delay performance for both systems is similar to that of wired Ethernet for an interframe space in the region of 10 ms and above.

The detailed experimental results presented in this chapter will be useful for wireless LAN designers. The primary source of inaccuracy in the measured results is the need for packet transmission from a wired segment. A wireless network analyzer may provide a more accurate evaluation since a direct wireless transmission incurs less delay compared to an indirect transmission from the wired network. However, such an analyzer is unlikely to appear commercially since different wireless trans-ceiver cards (for different physical layer wireless LANs) are required. The performance evaluation method presented in this chapter provides accu-rate results since the additional delay due to the wired transmission is insignificant (Ethernet packets are generated at the MAC sublayer) and there is no contention on the wired segments (since there is only one analyzer in each segment). This claim is substantiated by the measured throughput results which show a close agreement with the performance results supplied by the manufacturers. Furthermore, since wireless LAN products are likely to include wireless bridges for easy integration with wired networks, the technique of using network analyzers for quantita-tive performance evaluation of wireless LANs can be widely applied.

References

[1] Cali, F., M. Conti and E. Gregori, "IEEE 802.11 Wireless LAN: Capacity Analysis and Protocol Enhancement," *Proceedings of IEEE INFOCOM*, March 1998, pp. 142–149.

[2] Chhaya, H., and S. Gupta, "Performance of Asynchronous Data Transfer Methods of IEEE 802.11 MAC Protocol," *IEEE Personal Communications Magazine*, Vol. 3, No. 5, October 1996, pp. 8–15.

[3] Crow, B., et al., "IEEE 802.11 Wireless Local Area Networks," *IEEE Communications Magazine*, Vol. 35, No. 9, September 1997, pp. 116–126.

[4] Duchamp, D., and N. Reynolds, "Measured Performance of a Wireless LAN," *Proceedings of the IEEE Local Computer Network Conference*, September 1992, pp. 494–499.

[5] Bing B., and R. Subramanian, "A Novel Technique for Quantitative Performance Evaluation of Wireless LANs," *Computer Communications*, Vol. 21, No. 9, July 1998, pp. 833–838.

[6] Lucent Technologies, *User's Guide to WavePOINT-II*, November 1998.

[7] Symbol Technologies, *Spectrum24 Ethernet Access Bridge-User Guide*, August 1998.

[8] Claessen, A., L. Monteban, and H. Moeland, "The AT&T GIS WaveLAN Air Interface and Protocol Stack," *Proceedings of the 5th IEEE International Symposium on Personal, Indoor and Mobile Radio Communications*, September 1994, pp. 1442–1446.

[9] Bing, B. "Measured Performance of the IEEE 802.11 Wireless LAN," *Proceedings of the 24th IEEE Conference on Local Computer Networks*, Lowell/Boston, MA, October 1999.

[10] Boggs, D., J. Mogul and C. Kent, "Measured Capacity of an Ethernet: Myths and Reality," *Proceedings of ACM SIGCOMM*, September 1988, pp. 222–234.

[11] Shoch, J. and J. Hupp, "Measured Performance of an Ethernet Local Network," *Communications of the ACM*, December 1980, pp. 711–721.

[12] AT&T/NCR, *WaveLAN Design Guide*, 1994.

[13] Freeburg, T. "Enabling Technologies for Wireless In-Building Communications: Four Challenges, Four Solutions," *IEEE Communications Magazine*, Vol. 29, No. 4., April 1991, pp. 58–64.

[14] Buchholz, D., P. Odlyzko, M. Taylor and R. White, "Wireless In-Building Network Architecture and Protocols," *IEEE Network*, November 1991, Vol. 5, No. 6, pp. 31–38.

[15] Motorola Inc., *Altair Plus II System Manual*, 1992.

Contents

Wireless ATM Networks

Wireless ATM networks are wireless extensions to both public and private wired ATM networks. Unlike wireless LANs, wireless ATM will see a high demand in wireless mobile applications involving multimedia. For wireless ATM, although physical and link layer issues such as interference and multiple access assume a high degree of importance, it is user mobility and its implications for service provision that predominates in all mobile ATM systems. This chapter presents a coherent framework on the technical and service aspects related to the provision of broadband wireless ATM services.

6.1   ATM technology

The asynchronous transfer mode (ATM) is a data transport technology that supports a single high-speed infrastructure for integrated broadband communications involving voice, video, and data. ATM achieves bandwidth efficiency through statistical multiplexing (sharing) of transmission bandwidth. ATM networks are characterized by virtual circuit connections that carry short, fixed-length packets (called cells) within the network irrespective of the applications being supported. In setting up connections, the network makes resource allocation decisions and balances the traffic demands across network links, thereby separating data and control flows and enabling switches to be simpler and faster. At the network edge or at the end-equipment, an ATM adaptation layer (AAL) maps the services offered by the ATM network to the services required by the application. This enables ATM to handle a wide range of information data rates together with various types of real-time and nonreal-time service classes with different traffic attributes and quality of service (QoS) guarantees. Unlike LAN technologies, ATM is distance-independent and can be deployed in local or wide area networks [1]. As can be seen in the current ATM physical layer interfaces listed in Tables 6.1 and 6.2, ATM accommodates different data rates at a wide variety of distances. With wireless ATM, ATM technology will become both distance and physical medium independent.

6.1.1 Comparison of transfer modes

Figure 6.1 compares the operation of three common transfer modes, namely the synchronous, packet, and asynchronous transfer modes. The synchronous transfer mode employs the concept of a frame to map the

Table 6.1
Current ATM Physical Layer Interfaces for the Private Network

Data Rate	Physical Medium	Distance
25.6 Mbps, 51.84 Mbps, 155.52 Mbps	UTP-Category 3 (voice-grade)	100 m
155.52 Mbps	UTP-Category 5 (data-grade)	100 m
155.52 Mbps	STP	100 m
51.84 Mbps, 155.52 Mbps	Coaxial	2 km
100 Mbps, 51.84 Mbps, 155.52 Mbps	Multi-mode Fiber	2 km
622.08 Mbps	Multi-mode Fiber	300 m
51.84 Mbps, 155.52 Mbps, 622.08 Mbps	Single-mode Fiber	2 km

Table 6.2
Current ATM Physical Layer Interfaces for the Public Network

Data Rate	Physical Medium	Distance
1.544 Mbps	Twisted Pair	3000 ft
2.048 Mbps	Twisted Pair	Unspecified
$n \times$ **1.544 Mbps**	Twisted Pair	Under Study
2.048 Mbps, 6.312 Mbps, 34.368 Mbps	Coaxial	Unspecified
44.736 Mbps	Coaxial	900 ft
51.84 Mbps, 155.52 Mbps, 622.08 Mbps	Single-mode Fiber	15 km

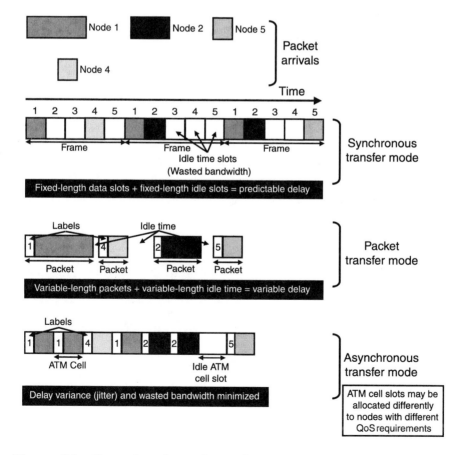

Figure 6.1 Operation of transfer modes.

position of a time slot to a node. Each frame is transmitted regularly, thereby guaranteeing each time slot (and hence each node) a fixed bandwidth. The advantage of such systems is that they can be implemented on hardware to provide fast bandwidth switching. Packet transfer makes use of an explicit label on the header of each packet to identify the destination node. These packets are independently multiplexed, buffered, routed, and forwarded on a link-by-link basis by various nodes of the network. In such an approach, packets are multiplexed and merged over the available paths on a statistically determined basis, gracefully adapting the transmission to different traffic levels and optimizing the use of existing link capacity without pre-allocating link bandwidth. In addition, two communicating entities have the possibility of utilizing all the available bandwidth of the link connecting the two nodes. However, the complexity of routing algorithms means that such systems are often implemented on software. ATM is a variant of the packet transfer mode with the key characteristic that all packets are short and of the same lengths (53 octets each). This is in contrast to the longer and variable-length packets found in packet networks such as X.25, Ethernet, and wireless LANs. The ATM packet structure reduces delay variation and simplifies the design of hardware-based and highly parallel packet switches. Moreover, when real-time traffic (e.g., voice, video) requires bandwidth, the bandwidth that has been allocated to less time critical traffic (e.g., data) can be held up quickly. By conveying data in short cells, ATM can rapidly interleave bursts of traffic from different connections, thus allowing network resources to be shared collectively among all connections. More significantly, this very fine-grained sharing enables the cost of a link to be divided among many users. In wireless ATM, where potentially 50,000 ATM cells can be multiplexed into the same 25 Mbps broadcast wireless link [2], statistical multiplexing gains can be considerable.

An equally important advantage of ATM is the flexibility to support connections of any desired capacity. This is achieved by changing the rate at which ATM cells are transmitted and by subdividing the overall network capacity into multiple virtual circuits, each carrying a variable number of multiplexed connections. The flexible data rates allow a wide range of services to be supported. Present-day telecommunication networks are typically implemented using a fixed hierarchy of link speeds.

For example, digitized voice circuits operate at 64 Kbps while the link speeds for T1 and T3 are about 1.5 Mbps and 45 Mbps respectively.

The characteristics of the three transfer modes and their abilities to support different traffic types are summarized in Tables 6.3 and 6.4 respectively. The effectiveness of ATM in transporting real-time traffic such as voice and video is primarily due to the short, fixed-length packet transmission which results in low packetization delay and low delay variability. Since multimedia traffic comprises a combination of two or more traffic types, ATM is the only transfer mode that can support multimedia traffic efficiently.

ATM also takes advantage of the pipelining effect in wide area networks to achieve good end-to-end delay performance (see Figure 6.2). Many ATM cells of the same data packet may be in transmission over consecutive links from the source node to the destination node, thus achieving a pipelining effect that considerably reduces end-to-end network delay. Even though the segmentation of a packet into shorter ATM cells increases the number of cells to be switched, inherently produces

Table 6.3
Characteristics of Transfer Modes

Characteristic	Synchronous	Packet	Asynchronous
Multiplexing	Position	Label	Label
Bandwidth Access	Periodic	Statistical	Statistical
Unit of Transfer	Fixed Length	Variable Length	Fixed Length
Bandwidth Allocation	Fixed	None	Flexible

Table 6.4
Effectiveness of Traffic Support by Transfer Modes

Traffic Type	Synchronous	Packet	Asynchronous
Voice Traffic	Excellent	Poor	Good
Video Traffic (Uncompressed)	Excellent	Poor	Good
Video Traffic (Compressed)	Good	Good	Excellent
Data Traffic	Poor	Excellent	Excellent

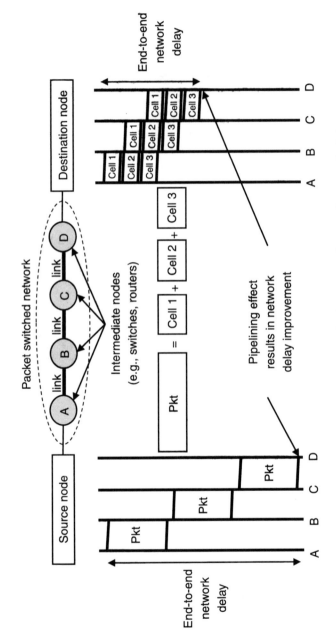

Figure 6.2 Effect of short-length packet transmission on end-to-end network delay.

queuing delay between cells of a packet, and introduces extra header overhead, these disadvantages are more than offset by the much larger gains due to the pipelining effect (see Figure 6.3).

Unlike other connection-oriented packet switching methods such as X.25, ATM has no link-by-link flow and error control. It is left to the applications at the network end-points to implement end-to-end flow and error control when required. By removing these functions, the processing time for each ATM cell is considerably reduced. In addition, minimum time is needed to switch an ATM cell because routing is performed only once (i.e., during connection setup). The use of virtual circuit connections in ATM implies that only relative addresses between different nodes of the network are required. Relative addressing incurs less overhead compared to global addressing and is suited for hardware-based rooting. The disadvantage of relative addressing is the need for a connection setup phase that links the global address with the relative addresses.

6.1.2 ATM versus IP

A fundamental performance issue with store-and-forward packet switching occurs when a long packet gets ahead of a short one (see Figure 6.1), particularly on a lower-speed wide area network connection. When short packets find themselves queuing behind longer packets, excessive delays may arise. Because long packets are more likely to have large variation in lengths, the delays encountered by short packets may also be highly variable. If an application is sensitive to delay or variation in delay, it may not be able to wait until the long packet completes transmission. For example, a 1,518-octet Ethernet packet takes approximately 8 ms to transmit on a 1.5 Mbps line. Intermediate nodes increase this delay further. Voice over IP implementations compensate the large delay variation by inserting a large playback buffer at the destination node. A simpler solution is to increase the link speed and/or decrease the maximum packet length like in ATM switching. Since long packets are segmented into shorter ATM cells, network delays can be smaller and significantly less variable compared to IP switching. The short, fixed size packets adopted by ATM also enable the construction of large, scalable and cost-effective switches. IP switches will have to deal with packets as large as several thousands of octets.

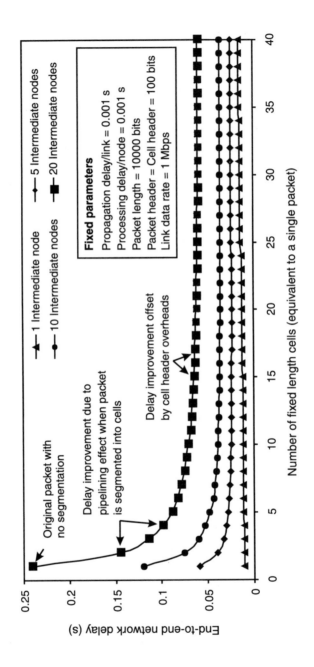

Figure 6.3   Effect of packet segmentation on end-to-end network delay.

ATM reserves bandwidth and distinguishes traffic characteristics in order to service connections with different QoS. This explicit control over QoS is in contrast to the single QoS offered by best effort IP (and telephone) networks. When there is excess bandwidth, both best effort and reserved bandwidth service work well. However, when there is a network congestion, best effort service slows down all users while reserved bandwidth service blocks some users in order to fully serve others. For applications involving real-time voice and video, a reserved bandwidth is preferable in many instances [3]. ATM allows bandwidth reservation to be performed using switched virtual connections. Unlike permanent virtual connections where the bandwidth is dedicated 100% of the time, switched virtual connections deliver bandwidth on demand, returning this resource when a node completes an activity or when the network node detects a period of inactivity. The connectionless nature of IP networks cannot support QoS in a manner that is superior to ATM.

With explicit bandwidth controls in ATM, a user can measure the allocated bandwidth. In contrast, because bandwidth is allocated without stringent enforcement in best effort IP service, there is no assurance that the service actually provided by the network is the one the user is supposed to get. Such feedback to users on actual network usage, even in the absence of usage billing, can encourage efficient and prudent use of network resources [3].

In an attempt to avoid the use of ATM entirely, protocols such as the reservation protocol (RSVP) have been proposed to provide dynamic QoS support needed for real-time services over IP networks. RSVP permits receivers to specify different QoS requirements. It is closely integrated with multicast services where receivers first determine the network path for senders to distribute traffic specifications and then distribute their own service requirements. The argument for RSVP is that options for QoS are needed but not as a replacement for best effort service since many applications cannot predict in advance what their bandwidth requirements will be. However, the key difference between ATM and RSVP when it comes to QoS is that RSVP only prioritizes packet delivery over a fixed bandwidth, whereas ATM guarantees bandwidth over switched virtual connections. Thus, although a key challenge for ATM is to achieve high connection setup rates and low connection setup times, about 40% of existing IP traffic still runs over ATM.

6.2 The need for wireless ATM

Besides ATM, another significant development in telecommunications emerged in the 1990s. The area of wireless personal communications, aided by the freedom to communicate anytime and anywhere, has become a phenomenal success for commercial cellular and paging services. For example, the burgeoning sale of mobile cellular phones has exceeded the number of personal computers sold worldwide. The subscriber base for wireless communications services is growing 15 times faster that the subscriber base for wired services and this pace is expected to accelerate in the coming years [4]. According to some forecasts, the number of mobile phone subscribers may reach one billion by the year 2010 and surpass fixed phone lines [5]. The phenomenal growth provides a clear demonstration of the significant value users place on portability as a service feature.

With cellular phones being readily available to the general public and increasingly relied upon, coupled with an expanding number of portable computing devices, ubiquitous nomadic access is likely to become dominant in the next century. This growth will occur in an environment characterized by rapid development of end-user applications and services towards the Internet and broadband multimedia delivery over the evolving fixed wireline infrastructure. Thus, it is apparent that a new generation of wireless networks will be needed for two main reasons:

1. To enable a wide range of wireless technologies to interwork seamlessly with existing wireline networks;

2. To meet the diverse traffic service and quality expected by current and future subscribers.

The convergence of ATM in the wireline and wireless domains serves as an effective platform in achieving these objectives. Even operators in the cellular radio and personal communication industries find ATM an efficient method for handling packetized voice communications while providing a potential to integrate data communications [6]. Consequently, the subject of wireless ATM has been the focus of active research in recent years. However, many open issues remain to be addressed and resolved. For instance, the combined concerns of wireless and multimedia communications do not yield simple solutions since link conditions and traffic load distributions fluctuate dynamically with time. In particular, ATM standards assumed a typical bit error rate (BER) of roughly 10^{-10} and an ATM cell loss ratio (CLR) of about 10^{-6}. These performance benchmarks are difficult to match with highly sensitive wireless communication links. The need to accommodate mobility while satisfying established QoS presents another serious problem since mobile nodes do not have permanent access points to the fixed network. On a broader perspective, the efficient support of multimedia information services will require a major shift from the current fixed-rate circuit-switched wireless mobile infrastructure towards integrated services packet/cell switching architectures. It is interesting to note that the performance of conventional ATM networks based on bandwidth optical fibers are limited by the switching capacity of the ATM switches. However, with wireless ATM, the performance bottleneck has now shifted from the switching capacity of the switches to the transmission bandwidth of the wireless link.

6.3 Wireless communications using ATM

Like wireless LANs, wireless ATM networks will have to deal with the time-varying nature of the wireless link. Multipath propagation, shadow fading, presence of physical objects along a wireless link (e.g., hills, buildings, vehicles, people, trees), and interference all affect the quality of the wireless link. These impairments vary with time and affect the network topology, link performance, and QoS delivered to the node. Such problems are virtually absent in conventional wireline ATM networks.

In a wireless environment, short packet transmission is robust against high fading rates (fast fading) that occur at high frequencies and high mobility speeds. For a wireless ATM data rate of 25 Mbps, the transmission time for a 53-octet ATM cell is roughly in the region of 17 µs. To transmit a maximum-length Ethernet packet (1,518 octets) at the same speed requires 500 µs, some 30 times higher than an ATM cell transmission. This means that the probability of a signal fade corrupting a wireless ATM cell is much lower compared to a standard data packet. As illustrated in Figure 6.4, a large percentage of ATM cells can be transmitted without

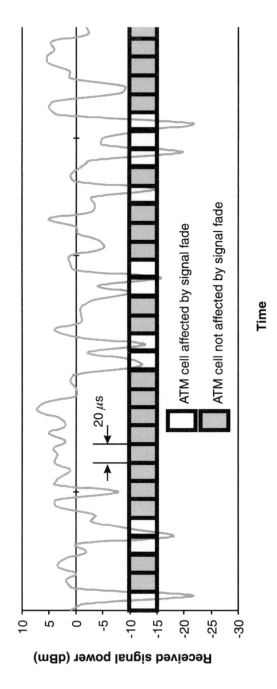

Figure 6.4   Effect of fading on short-length packet transmission.

error even as the signal fades fairly rapidly. Clearly, the retransmission of short ATM cells is more efficient, since the probability of a successful ATM cell retransmission is much higher compared to the retransmission of a long packet.

6.3.1 Wireless ATM transport modes

Although sending short ATM cells in isolation offers more protection against signal fading, this method can be inefficient since the time needed to transmit an ATM cell is typically less than the time needed to access a wireless link. Therefore, encapsulating ATM cells in longer packets before sending them will increase the transmission efficiency at the expense of reducing transmission reliability. To achieve this, a node will need to buffer several ATM cells until the required packet length is met, a process that increases the overall delay of the system. On the other hand, given the high data rates in wireless ATM, transmission of individual ATM cells can result in potentially large statistical gains.

Given these two arguments, wireless ATM cells can be transported using either the encapsulated or the native mode. In the encapsulated mode, ATM cells are grouped together and then carried over using a common wireless MAC protocol such as CSMA/CA. In the native mode, individual ATM cells are transmitted. The encapsulated mode has additional disadvantages such as protocol conversion at the wireless/wired interface, nontransparancy to ATM protocols, and lack of QoS support. Transmitting ATM cells using the native mode is disadvantaged by the high wireless overhead necessary for each cell.

6.3.2 Cellular architecture

Due to the high capacity requirement and the need to utilize the limited frequency spectrum efficiently, terrestrial wireless ATM networks that intend to cover reasonable distances must be built in a group of small geographical coverage zones called micro- or pico-cells. Like mobile phone cellular networks, the frequency channels of these coverage areas can be reused concurrently in distant locations (see Figure 6.5). The limited range of micro- and pico-cellular networks enables the implementation of low-power wideband devices supporting high-speed multimedia applications. These networks also provide accurate location

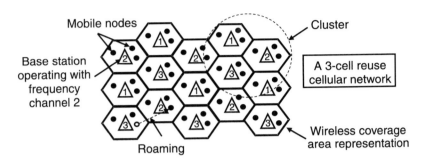

Figure 6.5 Frequency reuse in cellular networks.

information of mobile users. Because bandwidth is shared and spatially reused by many nodes, it is possible for two or more nodes to interfere with each other by accessing the same frequency band in different locations, thus giving rise to co-channel interference.

Irrespective of the multiple access technique employed, spectral efficiency for cellular networks is maximized when the system is interference limited. This means that while link budgets are important for determining coverage areas and power levels between specific transmitters and receivers, it is the interference from co-channel transmitters that ultimately limits the capacity and performance of the system. If the system is not interference-dominated, then the spectral efficiency can be further increased by allowing more mobile nodes or by reusing frequencies at shorter distances [7]. However, reducing the size of the coverage area increases the handoff rate (due to increased crossings between coverage areas per unit of time), which in turn increases the likelihood of dropped connections. In addition, routing becomes more dynamic because routes need to be re-established whenever a handoff occurs. This places great demands on the switching architecture.

A more severe problem arises in cellular networks when certain wireless coverage areas encounter significantly more traffic than others and service quality in these congested areas becomes adversely affected. This traffic imbalance issue is rather complex because it varies dynamically with time and is dependent on several factors such as the size of the coverage areas, the user density per coverage area, and the degree of mobility of the users. The situation can be improved if suitable traffic load balancing (management) schemes are implemented so that traffic from one wireless coverage area can be directed to another when desired [8].

6.3.3 Operating frequency

For wireless ATM networks, the choice of a frequency band is closely related to the propagation characteristics of signals in that band and the system requirements in terms of bandwidth and connectivity. Although the 2.4 GHz ISM band with 83.5 MHz of bandwidth can theoretically support data rates of up to a few tens of Mbps, this band is congested because it is being used by IEEE 802.11 wireless LANs and microwave ovens. The 5 GHz band is more attractive since it provides about 300 MHz of relatively clean spectrum in which only high-speed data systems and a few satellite systems actively operate. Although the 5 GHz frequency range seemed to be the predominant choice for wireless ATM, the quest for higher data rates may require higher frequencies in the 17 GHz and 60 GHz bands to be considered as future options. The 60 GHz band, for instance, seemed to be suitable for multimedia applications because enough bandwidth is available. At 60 GHz frequencies, propagation loss due to oxygen absorption is so high that reflected signals do not pose a significant problem. This also reduces co-channel interference and facilitates frequency reuse at short distances, thereby allowing the use of low-power transmitters. The FCC has opened 5 GHz of spectrum in the 59 to 64 GHz frequency band.

6.3.4 Modulation

The harsh physical layer characteristics of the wireless link drive the need to develop modulation schemes that are spectrally power-efficient and can mitigate the effects of interference and multipath fading. Modulation is the process of encoding information into the amplitude, phase, and/or frequency of a transmitted signal and this affects the bandwidth of the transmitted signal and its robustness under impaired channel conditions. Modulation techniques fall into two categories: linear and nonlinear. Linear modulation techniques consume less bandwidth but produce large fluctuation in signal amplitude that are easily distorted by nonlinear (but cheap and power-efficient) amplifiers. This implies that the bandwidth efficiency of linear modulation techniques is generally obtained at the expense of additional power needed for very linear amplifiers. Therefore, modulation techniques that have constant or nearly constant envelopes versus time are required due to power amplifier considerations. The constant envelope requirement is easily met by employing phase and frequency modulation schemes such as phase shift keying (PSK) and frequency shift keying (FSK).

In addition to efficient power amputation, good communications efficiencies in terms of low error probability for a given signal-to-noise ratio and good bandwidth efficiencies in terms of the number of bits per second per Hz of bandwidth are desirable. For a given bit error rate, the larger the number of bits encoded per symbol, the more efficient the use of bandwidth but the greater the power requirement.

For simultaneous good bandwidth and power efficiencies, combined modulation and encoding techniques can be used at the expense of increased implementation complexity. An example would be multilevel continuous phase modulation (CPM). Other possible modulation candidates for use in wireless ATM systems include differential quadrature phase shift keying (DQPSK), Gaussian minimum shift keying (GMSK), and multilevel quadrature amplitude modulation (QAM). Advanced receiver technologies such as adaptive equalizers and smart antenna arrays may have to be incorporated.

DQPSK can simplify receiver design because no absolute phase reference is required for demodulation. However, a known drawback of differential phase modulation is the sensitivity to receiver carrier frequency offset. This drawback can be resolved by using stable crystals to improve the accuracy of the carrier frequency at the receiver [9]. A modified version of DQPSK with differential coherent detection (known as $\pi/4$-shifted DQPSK) is popularly adopted by mobile cellular networks. The modification consists of rotating every second symbol by $\pi/4$ radians or 45 degrees. During odd time intervals, the phase values 0, $\pi/2$, and $3\pi/2$ are used to transmit information while in even time intervals, the phases $\pi/4$, $3\pi/4$, $5\pi/4$, and $7\pi/4$ are used. This modulation technique is essentially one form of minimum shift keying (MSK) since it reduces the amplitude or envelope variation of the modulated signal, thereby increasing the efficiency of the power amplifier. GMSK is also a constant envelope modulation scheme that does not require the use of a linear amplifier. This leads to a highly efficient power amplification compared to $\pi/4$-DQPSK (which is more spectrally efficient). The major difference between GMSK and $\pi/4$-DQPSK is that GMSK has a constant envelope whereas $\pi/4$-DQPSK has a constant envelope only at the sampling instants. A very desirable property of MSK and GMSK is that although they have constant amplitude, they can be implemented using linear modulation techniques based on quadrature-type architecture [10].

6.3.5 Equalization and antenna techniques

If omnidirectional antennas are used and high data rates need to be supported, then some form of equalization is normally unavoidable [11]. Equalization increases the complexity and power consumption of the receiver. Of major concern is the overhead introduced by training symbols that needs to be transmitted with each packet. This overhead makes it highly inefficient to transmit short ATM cells individually since the training period becomes long relative to the cell length. The use of directional antennas is an alternative to equalization. Switched-beam smart antennas can be used where each antenna comprises an array of beams which can be switched electronically at a rate sufficiently fast to follow the rate of change of the wireless link. The physical size of the antenna elements is inversely proportional to the operating frequency. This means that for higher frequencies, more elements can be accommodated and as a result, beam patterns can be more precise. Each element is controlled electronically through changes in the dielectric properties of the dielectric materials. Since the antenna does not change physically, no moving parts are required. Such steerable antennas can change the shape and direction of their transmit beams depending on the location of the mobile node. In doing so, a wireless coverage area can be broken down into smaller areas called sectors, each serviced by a directional antenna (see also Section 5.4). This is equivalent to increasing the number of communication links. Smart antennas, particularly when deployed at the base station, can also help to avoid or minimize the effects of multipath propagation and co-channel interference. This method, however, requires the signal to be locked on and tracked, functions that introduce additional overhead.

6.3.6 Multicarrier schemes

Multicarrier (or multitone) systems are parallel transmission schemes that compensate for the multipath delay spread without the need for equalization. The frequency band is divided into a number of subchannels, each modulated at a much lower symbol rate than a single carrier scheme using the same bandwidth. Because each subchannel is narrow enough to cause only flat fading, this makes a multicarrier system less susceptible to intersymbol interference. By adding a small guard interval, such interference can be completely eliminated. Like multilevel schemes, a major drawback of multicarrier systems is the high peak-to-average power ratio. For N carriers, if the peak power is limited, then the average power that can be transmitted is reduced by $1/N$. More importantly, the presence of envelope variation requires linear amplifiers which have lower power conversion efficiencies than nonlinear amplifiers.

An orthogonal frequency division multiplexing (OFDM) system uses less bandwidth than an equivalent multicarrier system because the frequency subchannels are overlapped. OFDM offers frequency diversity which can be exploited by proper coding to combat frequency-selective fading. In order to recover the information, it is necessary to use orthogonal signals in the subchannels. These signals can be generated and decoded using Fast Fourier Transform (FFT), which can be adapted to different data rates and different link conditions. Because both transmitter and receiver modulation can be achieved using FFT, this allows efficient digital signal processing implementation. On the negative side, OFDM is more sensitive to frequency offset and timing mismatch than single-carrier systems. The need to amplify a set of frequency carriers simultaneously mandates the need for low-efficiency linear power amplifiers, which leads to high power consumption.

Multicarrier techniques may also be applied to high-speed wireless optical systems since the high symbol rate using intensity modulation makes it hard to alleviate intersymbol interference. A further advantage of multicarrier optical modulation is that it moves the signal spectrum away from dc, thereby avoiding low-frequency harmonics due to fluorescent light interference [12]. An alternative to using multicarrier is multiwavelength modulation. In this case, each data substream modulates a different, noninterfering wavelength rather than a different carrier. Although this approach is appealing, it requires an expensive bank of narrowband optical filters. Semiconductor laser diodes are preferable over light-emitting diodes as an optical source for high-speed transmission because they have a higher optical power output capability and because their electrical-to-optical conversion is more linear.

6.3.7 Duplexing

Duplexing concerns the manner in which base-to-mobile (downlink) and the mobile-to base (uplink) transmission is conducted. The commonly used duplexing methods are frequency-division duplexing (FDD) and time-division duplexing (TDD). In FDD, the downlink and uplink transmission occurs simultaneously on different frequency bands. In TDD, the uplink and downlink transmission occurs in different time slots. The amount of spectrum required for both FDD and TDD is similar. The difference lies in that FDD employs two bands of spectrum separated by a certain minimum bandwidth (guard band), while TDD requires only one band of frequencies. TDD's strength lies in the fact that it may be easier to find a single band of unassigned frequencies than to find two bands of unassigned frequencies that match the required bandwidth. Since bandwidth can be allocated between uplink and downlink transmission in a flexible manner, TDD systems may be more suitable for multimedia applications with both symmetric and asymmetric connections. A voice call is an example of symmetric connection while video browsing is considered an asymmetric connection that involves sending control information in one direction and video transfer in the reverse direction. Another prime example is Internet traffic where downlink traffic predominates.

On the other hand, interference in FDD systems can be better managed since an uplink transmission can only interfere with another uplink transmission in the same frequency band. In TDD systems, a high-powered downlink transmission can interfere with a low-powered uplink transmission if the entire cellular network is not time synchronous. However, receiver design for TDD systems is simpler.

6.3.8 Multiple access

The weakness of distributed wireless LAN protocols such as CSMA/CA is that medium access has to be negotiated for each packet transmitted. Furthermore, carrier sensing reduces transmission efficiency due to the turnaround time in half-duplex transceivers. A multiple access protocol supporting wireless ATM must guarantee QoS, and enable many nodes in a wireless coverage area to share the same communications bandwidth efficiently and provide timely access to multirate broadband applications involving integrated voice, video, and data services. A survey of wireless ATM MAC schemes can be found in [13, 14].

6.3.9 Code division multiple access

The multiple access technique in spread spectrum and code division multiple access (CDMA) networks usually refers to the ability of certain kinds of signals to coexist in the same frequency and time space with an acceptable level of mutual interference [15]. The use of pseudorandom or pseudonoise waveforms in a wireless network is motivated largely by the desire to achieve good performance in fading multipath links and the ability to operate multiple links with pseudo-orthogonal waveforms using spread spectrum multiple access. Although the technology is promising, it is not clear whether spread spectrum-based systems can efficiently accommodate the high and variable data rates demanded by multimedia services.

Like spread-spectrum radio LANs, CDMA networks can be broadly classified under direct-sequence and frequency-hopping. A number of hybrid CDMA and multicarrier schemes [16] have been proposed for wireless broadband communications. These schemes are depicted in Figure 6.6. Multicarrier schemes employ parallel signaling methods that offer several advantages over conventional single carrier systems such as protection against dispersive multipath links and frequency-selective fading. With appropriate signal processing and forward error correction coding, these systems can achieve the equivalent capacity and delay performance of single carrier systems without the need for a continuous frequency band. Multicarrier schemes, when used in conjunction with OFDM, are bandwidth-efficient since guard bands between adjacent carriers are unnecessary in OFDM.

Recently, there has been an interest in providing high-speed CDMA transmission using code aggregation. A new version of IS-95B assigns up to eight codes. A similar approach involves multirate code division where each node accesses the link using a spreading factor equal to the available bandwidth divided by the required peak data rate. Other approaches for improving the performance of CDMA systems include interference suppression and interference cancellation. Interference suppression using adaptive signal processing may have an advantage over interference

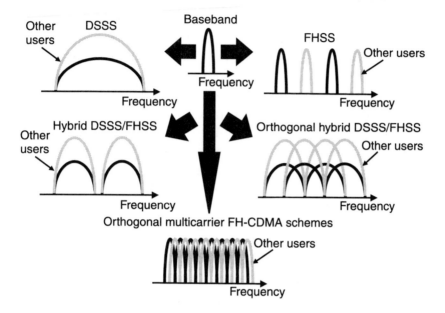

Figure 6.6 Spread spectrum multiple access techniques.

cancellation systems since it eliminates the need for accurate parameter estimation (e.g., received signal's amplitude and phase). In these methods, acquisition considerations (e.g., employing optimal and suboptimal receiver structures) and the use of adaptive spatial arrays play central roles.

Multiuser detection is another method that can improve the performance of CDMA systems. However, the simultaneous detection of multiple signals requires low bit error rates because bits that are erroneously detected are subtracted from the signals of other users, potentially causing those signals to be decoded in error as well.

6.3.10 Time division multiple access and polling schemes

Many multiple access schemes currently deployed in mobile cellular networks employ a fixed frame structure (e.g., time division multiple access or TDMA). Such schemes are optimized for continuous traffic rather than bursty traffic. To cater for packetized multimedia traffic, the selection of a suitable frame structure is not an easy task since there is little knowledge of the traffic mix [17]. If time slots (that constitute a frame) are chosen to match the largest packet lengths (each packet comprising several ATM cells), slot times that are under-utilized by short packets must be padded out to fill up the slots. On the other hand, if shorter slot sizes are used, more overhead per packet results [18]. For some TDMA schemes, long frames are necessary in order to maximize the bandwidth utilization at high traffic loads but this is done at the expense of increasing the delay at low loads.

An alternative approach abandons the concept of a frame reference altogether. Instead of choosing a basic terminal bit-rate to fix the duration of a slot as in TDMA, a hybrid multiple access strategy described by Bing and Subramanian [8] achieves more flexible bandwidth sharing by allowing mobile nodes to seize variable amounts of bandwidth on demand. To provide QoS guarantees, the scheme adapts to changes in traffic load, interference conditions, and user mobility autonomously. Unlike other random access protocols, it can be shown that this contention-based multiple access technique achieves stable throughput [19].

6.3.11   The simple asynchronous multiple access protocol

A multiple access protocol known as simple asynchronous multiple access (SAMA) has been proposed for wireless ATM [20]. This protocol provides a simple bandwidth setup mechanism with support for widely varying data rate requirements. The protocol can be slotted or unslotted. The operation of slotted SAMA is shown in Figure 6.7. Bandwidth allocation is depicted in Figure 6.8.

6.3.12   Error control

In wireline ATM, there is no specified mechanism for error recovery or retransmission at the link layer. In wireless ATM, because of the higher error rate of the wireless link, some of the QoS issues need to be addressed at the link layer rather than from an end-to-end basis. To this end, a combination of forward error correction (FEC) coding and automatic repeat request (ARQ) is effective in improving QoS parameters such as the CLR.

FEC techniques typically use error correcting codes (e.g., convolutional coding, block coding) that can detect with high probability, the error location. Convolutional codes encode a continuous sequence of bits.

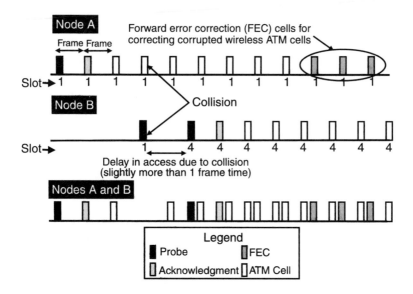

Figure 6.7 Slotted simple asynchronous multiple access.

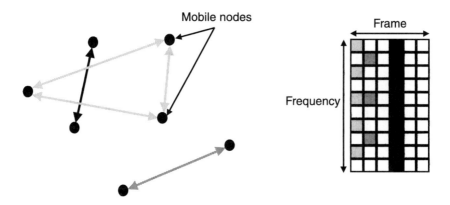

Figure 6.8 Bandwidth allocation in SAMA.

Advanced schemes such as Trellis codes combine channel coding and modulation while Turbo codes exhibit superior error correction performance based on iterative and parallel decoding of two concatenated convolutional codes but are generally very complex and produce large delays. These channel codes improve bit error rate performance by adding redundant bits in the transmitted bit stream that are employed by the receiver to correct errors introduced by the channel (medium). Such an approach reduces the signal transmit power for a given bit error rate at the expense of additional overhead or a reduced data rate (even when there are no errors). Channel codes are best suited for correcting 1 or 2 contiguous bit errors in relatively benign transmission links. The performance of such codes deteriorates rapidly when errors occur in large bursts.

Due to the bursty nature of errors encountered in a fading link, block coding (e.g., Bose-Chaudhuri-Hocquenghem codes, concatenated Reed Solomon codes) is preferred over convolutional coding. Block codes add parity bits to blocks of messages. Besides error correcting capability, these codes also have an excellent error detection capability which can be used in conjunction with the ARQ method. Interleaving to combat burst errors caused by deep signal fades may be necessary. Interleaving results in time diversity because the technique converts burst errors to random errors that can be corrected easily. However, the increased delay of coding and interleaving can lead to a loss of spectral efficiency when the fading rate is slow relative to the data rate.

Employing FEC alone is not effective since many ATM cells may be lost or dropped due to buffer overflow at the receiver. In ARQ, the receiver employs error detection codes to detect errors in the received packets and then requests the transmitter to resend any error packet. It is simple to implement and is most useful when the link characteristics are unknown or unpredictable. However, the ARQ technique improves the CLR at the expense of lowering the cell delay variation (CDV) performance. User interactivity requirements impose a limit to the number of ARQ retransmissions. In general, applications that are not delay-sensitive can be transmitted with ARQ while real-time applications involving voice and video are better transmitted with FEC.

A typical wireless ATM cell structure is shown in Figure 6.9. A cell sequence number is added to each ATM cell to identify each cell uniquely for acknowledgment and retransmission purposes. A cyclic redundancy check (CRC) code is appended at the end of each ATM cell for the detection of errors. These two fields are valid only on the wireless link and are stripped off before the ATM cell enters the ATM layer. Note that the large overhead in ATM headers (about 10%) will result in a reduction in transmission efficiency. Extra overhead due to wireless headers/trailers and wireless operation (e.g., equalizer training, antenna adaptation, etc.) reduces this efficiency even further. In order to improve the efficiency of wireless transmission, the header of the standard ATM cell

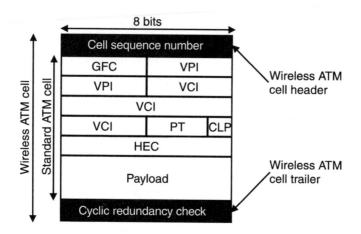

Figure 6.9 A typical wireless ATM cell format.

may be compressed to between 2 to 4 octets. The header error control (HEC) may also be removed by the transmitting node and then re-inserted after the wireless ATM cell has been successfully received.

It should be pointed out that the use of FEC together with the HEC in the ATM cell forms a concatenated code that is not optimized. The problem is that a FEC decoder will produce burst errors at the output whenever decoding errors occur. This will in turn produce an uncorrectable error in the HEC if the error burst occurs within the cell header. The HEC function performs poorly here because it is designed to detect and correct random (and not burst) errors [21].

6.3.13 Transport control

As wireless networks have much higher error rates, lower bandwidth, and more frequent outages than wireline networks, wireline network protocols cannot provide adequate functionality or performance. For example, the performance of conventional transport control protocol (TCP) over a wireless link is poor because TCP's congestion avoidance algorithm (which is controlled by the sender) works best only for networks that experience low packet loss. TCP makes the implicit assumption that retransmission is a result of network congestion. Thus, pauses during handoffs (when users move among different wireless coverage areas) can be perceived as periods of heavy losses by the transport layer, causing retransmission timeouts. Packet losses over the wireless link can also lead to retransmission timeouts. In both cases, TCP reacts by drastically reducing the current transmission rate [22]. First, TCP reduces the transmission window size to restrict the amount of data flowing through the network. Second, it activates the slow-start mechanism that lowers the incremental rate of the window size to only one packet. Finally, TCP resets the retransmission timer to a backoff interval that doubles with each consecutive timeout. These measures reduce the load on intermediate links, thereby controlling congestion on the network. However, TCP takes a long time to recover from a transmission rate reduction, resulting in severe throughput degradation. Such problems can be mitigated through a TCP-aware link layer in which the base station triggers local retransmission [23]. This approach attempts to make the lossy wireless link appear as a higher quality link with a reduced effective bandwidth. As a result, most of the losses seen by the TCP sender are caused by congestion. Another method attempts to make the sender aware of the existence of wireless links and realize that some packet losses are not due to congestion. The sender can then avoid invoking congestion control on noncongestion-related losses.

6.3.14 Resource allocation

In the fixed network infrastructure where interactive multimedia services are already experiencing remarkable growth, the provision of such services to mobile nodes will have to be supported. The extension of multimedia services over multiple operating environments (e.g., wireline and wireless ATM networks), characterized by widely different features (e.g., bandwidth and signal quality), calls for differentiated QoS provision to be dynamically maintained and adjusted according to the network and link operating conditions. Strategies are needed to both reduce the QoS variability and to conceal QoS impairments from the supported applications. In particular, the base stations of cellular wireless ATM networks will need to provide assurance that QoS requirements will be met. This can be achieved by explicit resource allocation using a combination of admission control, traffic shaping, and policing mechanisms. Requests for new connections are blocked if the anticipated traffic load presented by a new connection causes unacceptable congestion to build for existing connections. The connection admission mechanism must also insure a low rate of dropped connections as users roam among different wireless coverage areas. The admission decision is usually based on several criteria such as:

> Traffic and handoff characteristics;

> Call holding time statistics;

> Desired QoS of each class of traffic;

> Amount of radio resource available.

Many rate-based scheduling schemes have been proposed for scheduling heterogeneous traffic whereby each traffic stream is assigned a guaranteed rate of service. This approach induces an undesirable coupling between bandwidth and latency requirements [24]. As an example, some interactive applications may have low latency and low bandwidth requirements. In order to satisfy the low latency requirement, such applications are usually guaranteed a rate greater than its bandwidth requirements, resulting in inefficient utilization of resources. A more efficient strategy is dynamic resource allocation in which bandwidth and power levels are assigned depending on the current interference, propagation, and traffic conditions. If absolute guarantees are too costly, then it may be sufficient to provide predictive service, indicating that the application's requirements are most likely to be met.

6.3.15 Mobility management

Mobility imposes many constraints and poses a significant challenge to current ATM protocols because the protocols are designed primarily for fixed terminals. Mobility management refers to personal and terminal roaming issues such as handoff signaling, paging, location registration, location update, authentication, and connection control (see Figure 6.10). The main mobility management functions are listed in Table 6.5.

Regardless of whether nodes are mobile or stationary, communication networks must perform basic functions such as controlling access to services, locating nodes, and routing traffic. Since a mobile node is usually located in a wireless coverage area on a temporary basis, all of these functions must be performed more frequently and at higher speeds. The connection-oriented transmission of ATM operates based on an initial negotiation with the network for a virtual circuit (VC), each characterized

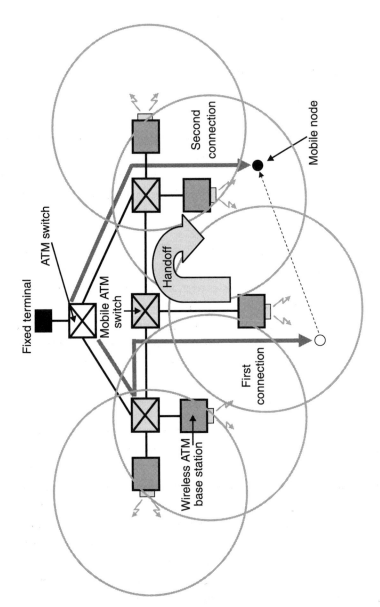

Figure 6.10 Mobility management and handoff.

Table 6.5
Mobility Management Functions

Mobility Function	Description
Location Management	Responsible for finding the mobile node.
Handoff Signaling	Refers to the process of changing frequency channels so that uninterrupted service can be maintained when nodes move across wireless coverage areas. This process also helps to track the mobile node dynamically.
Connection Control	Deals with connection routing and QoS maintenance.

by some allowable traffic profile and QoS. In a mobile environment, management of the VC with QoS is not easy since the end-to-end path has to be continually modified as terminals move during the lifetime of a connection. In some cases, the mobile user may move from one wireless coverage area to another between connection request and establishment. Because wireless base stations and mobile routing nodes are normally less capable than the wireline network counterparts, wireless ATM networks may potentially suffer from excessive delay and latency. For example, a change in routed path may result in jittered, lost, wrongly sequenced, or in some cases, duplicate messages, impairments that may lead to annoying service interruptions. In addition, the allocation of resources have to be re-evaluated each time a node moves to a new location. The situation becomes worse when the size of each wireless coverage area reduces to accommodate greater capacity per unit area. Moreover, mobile routing protocols need to operate in both wireline and wireless environments if they are to be usefully integrated into future networks. Hence, the routing of ATM cells to mobile terminals requires new mechanisms.

One method is to assign resources as a function of user mobility. A fixed or low-mobility node may be assigned higher bandwidth for a longer duration. In the virtual-connection tree method, the admission controller guarantees that the QoS metric negotiated during connection set up is maintained when a user roams among a limited set of wireless coverage areas. This is achieved by assigning virtual connection numbers to the path from the root to the leaves of the tree. Although mobile IP's tunnel-based triangle routing can also be considered, this approach is susceptible to breaches of security. Unless measures are taken to authenticate the mobile host, it may easily masquerade as another host or worse, hijack a legitimately authenticated session since the foreign agent welcomes a wide variety of hosts [25]. In addition, the practice of filtering source addresses as commonly adopted by firewalls to keep out unwanted intruders can be incompatible with the tunneling routing algorithm.

Developing solutions that ensure QoS resources keep pace with continually changing network states resulting from user mobility, without consuming large amounts of overhead in the process, will continue to be a fertile ground for wireless ATM research. More information on this subject can be found in [26].

6.4 Multimedia communications using wireless ATM

Multimedia remains an integral component of all evolving telecommunication systems. Applications such as video on demand, multimedia Internet access and messaging, collaborative multiparty teleconferencing, virtual retailing, and digital libraries are all influenced by the emergence of multimedia. Multimedia communications can be defined as the means of transferring information via voice, video, audio, image, graphics, text, or any combination of these media. The information content of one medium may affect the information generated by another medium. Multimedia communications provide the means to manipulate, transmit, and control multimedia information. Since multimedia applications are the most general class of applications, they have the most wide-ranging traffic attributes and communication needs. These applications impose various performance requirements on the network. The requirements are expressed in terms of QoS parameters which are based on traffic-dependent performance metrics such as allowable bandwidth, maximum delay, delay variation (jitter), and error rates. Bandwidth must be guaranteed for an application not only to satisfy the bandwidth requirement but also to limit the delay and errors introduced.

A multimedia network has to support a broad range of data rates not only because there are many communication media but also because the communication medium may be encoded by algorithms that generate different data rates. Efficient video coding schemes (such as those adopting the MPEG algorithm) that can maintain the same quality for all images inherently produce variable bit rate (VBR) outputs. In circuit-switched networks, coding schemes must reduce the resolution during complex scenes to achieve the constant bit rate (CBR) requirement [27]. As a result, constant image quality cannot be maintained for all scenes. In ATM networks, where bandwidth can be allocated flexibly on demand and statistical multiplexing is used, VBR coding schemes can be used without sacrificing bandwidth utilization efficiency. Figure 6.11 shows how a bursty VBR video traffic source can be shaped (adapted) to a CBR traffic stream using a smoothing buffer. Since each video frame is compressed with a different ratio, the frame boundaries after shaping become irregular. However, because the average bit rate of the shaped VBR video stream is now lower than the peak rate of the uncompressed CBR video stream, the number of simultaneous users can be increased.

The need for integrated multimedia communications requires a good understanding of multimedia traffic characteristics. In general, multimedia traffic can be categorized under periodic and bursty traffic types [28]. Message generation that occurs at regular intervals is an example of periodic traffic pattern. All real-time applications requiring time-based information to be presented at specific instants have periodic traffic patterns. An example is 64 Kbps PCM-digitized audio which produces

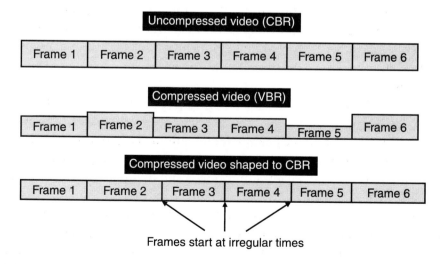

Figure 6.11 Traffic shaping.

samples at 125 μs intervals, each sample comprising 8 bits. In the case of uncompressed full-motion NTSC video, video frames (each containing a fixed amount of information) are generated at regular intervals of 1/30th second (or 30 frame/s). Even for compressed video (e.g., MPEG), a video frame is still created at regular intervals of 1/30th second for NTSC and 1/25th second for PAL formats except that the amount of information generated is variable at each instant depending on the degree of compression for each frame. These applications have strict timing bounds because a maximum delay is allowable after which the information generated is no longer useful and can be discarded. On the other hand, early arrivals can lead to buffer overflow because previous information has not been played out and are still in the buffer. Hence, the high delay variation introduced by real-time streaming applications can lead to high loss.

Bursty traffic is characterized by messages of arbitrary lengths that are generated at random time instants and separated by interframe intervals (idle time) of random duration, often resulting in a high peak-to-average data rate ratio. Hence, to meet the needs of new multimedia applications such as image/video browsing that have much burstiness is even more demanding.

The ATM Forum's traffic management specification (version 4.0) defines five ATM layer service categories: CBR, real-time VBR, nonreal-time VBR, available bit rate (ABR), and unspecified bit rate (UBR). ABR and UBR are best-effort traffic classes. ABR guarantees zero losses (but no other guarantees) if the source follows the traffic management signals delivered by the network. UBR provides no performance guarantees. QoS is measured using a set of parameters such as CLR, CDV, and maximum and mean cell transfer delay (CTD). More details can be found in Kwok and Toy [28, 29].

Generally, a wireless link is most efficiently shared among nodes with CBR requirements. However, the wireless network may only provide approximate performance guarantees to multimedia applications. Nevertheless, because multimedia transmission quality is usually based on user perception, multimedia communication is still useful at various levels of quality. Experience has shown that corrupted packets are more noticeable in an audio stream than in a video stream. For example, in a video conferencing application, people can communicate solely using voice but the video component alone may not be satisfactory. Similarly, real-time streaming applications involving audio or video streams are more tolerant of errors than data applications. Thus, unlike conventional data communications, occasional loss of voice or video packets may not necessarily result in devastating consequences. These observations imply that it is more important to increase the probability that most packets arrive by their expected deadlines than to achieve the successful arrival of all packets with no regard to delay.

Most existing video and image compression algorithms apply spatial and temporal interpolations based on reliably received information. For instance, the MPEG algorithm uses intraframe techniques that exploit the spatial redundancy within a picture as well as interframe techniques that take advantage of the temporal redundancy present in a video sequence. These schemes breakdown under high and pronounced fading effects typical of wireless links. On the other hand, because image and video compression algorithms normally produce information with different levels of importance, a higher degree of error protection can be accorded to more significant information while less important information can be discarded when link conditions are poor. For example, low resolution (base-level) portions of a compressed image is more vulnerable to error propagation (but more delay-tolerant) than portions that contain fine details. The segregation of important from less important data is best done by compression techniques. Each video frame can be compressed separately and independently from other frames so that the impact of transmission errors is restricted only to a single compressed frame. Such an intraframe approach is naturally less prone to errors compared to interframe coding where prediction and interpolation methods are used to achieve a high degree of compression between consecutive video frames. Differential protection for synchronization information is also vital since this affects the quality of video playback. Note that real-time interactive applications impose time limits on the compression process and this may in turn limit the quality of the compression.

Besides unequal levels of error protection, applications must also have the ability to adapt quickly to changes in the available bandwidth or network delay. If applications running on different user terminals are designed to automatically compensate for link impairments independently and without intervention from the source, graceful service degradation is ensured as the quality of the wireless link deteriorates. Such a function is particularly useful for the multicast of multimedia information where link errors and transmit power considerations limit

the effectiveness of sustained wireless broadcasting. Essential requirements to support adaptive applications include:

- Memory buffers at the receiver;

- Flow control if the receiver does not have sufficient buffer capacity
  to accommodate all the data received;

- Processors to handle intersample spacing and playback point at the
  receiver;

- Layered (hierarchical) coding, e.g., multiresolution (scalable)
  source coding;

- Error concealment techniques based on spatial or temporal interpolation from the adjacent areas of the same frame or the previous
  frame (these techniques require detection of packet loss in order to
  locate the damaged areas of the image).

Recent research investigated the adaptation of compression algorithms to changing link quality [8]. The motivation for such joint
source/channel coding came about due to the unpredictable and impre-
cise information provided by the wireless link. The overall capacity is
divided dynamically between the compression algorithm and channel
coding. Optimal performance can be achieved in the following manner.
If the quality of the link is good, all the capacity is allocated to the
compression scheme and no channel coding is needed. As the link quality
degrades, increasing levels of capacity allocated to channel coding to deal
with errors caused by the link. Such an integrated approach may have
implementation difficulties because source and channel coding techniques are traditionally developed separately.

6.5 Wireless ATM prototypes

Several wireless ATM proof-of-concept prototypes have been developed
by research laboratories around the world. They include WATMnet,
BAHAMA/MII, Magic WAND, AWACS, and others.

6.5.1 **WATMnet**

WATMnet is developed by the Computer and Communications (C&C) Research Laboratories of NEC in the United States. The components of the experimental system's hardware are laptop computers with wireless ATM network interface cards, multiple wireless ATM base stations, and an ATM switch with signaling extensions for mobility support. The wireless network interface cards operate at a maximum data rate of 8 Mbps in the 2.4 GHz ISM band.

The prototype employs TDMA with time division duplex and contention for multiple access. Bandwidth allocation is controlled by base stations with mobile terminals scheduling traffic on the allocated bandwidth. The data slots are partitioned into two groups:

1. Rate mode (for ABR, VBR, and CBR);

2. Burst mode (for UBR).

The protocol achieves 60% to 70% throughput efficiency with reasonable latency for a typical ATM traffic mix.

Two error-recovery modes are used for data link control, namely, zero loss mode and fixed recover window mode. In the zero loss mode, cell recovery is attempted without any limits on recovery latency. This mode is usually performed on nonreal-time services that do not have any delay concerns. In the fixed recover window mode, cell recovery is attempted within a fixed time window (selected at connection set-up). This mode is suitable for real-time traffic.

A custom wireless control protocol is implemented between the portable and base units for user registration and handoff. Dynamic rerouting of VCs is employed. This involves path extension from one radio port to another and/or re-establishing VC subpaths through new ATM switches or ports.

The field trial of a 25 Mbps enhanced WATMnet system operating in the 5 GHz band is expected soon. More information can be found in Raychaudhuri [30].

6.5.2 **BAHAMA/MII**

The wireless Broadband Ad Hoc ATM LAN (BAHAMA) is a self-organizing ad hoc network developed by Bell Laboratories. The system is designed for total portability in that both the node and the base station are portable. The portable base stations (PBSs) communicate to determine the topology of the network after changes due to the addition or deletion of other PBSs. To simplify the design of the PBS, ATM segmentation and reassembly are performed by the mobile terminals. To accommodate mobility, a simple virtual path identifier/virtual channel identifier (VPI/VCI) concept is defined which supports connectionless-type routing based on the destination address. Handoffs are executed using a novel homing algorithm that preserves ATM cell sequence within a connection, thereby maintaining established QoS. The network employs a wireless data link layer that ensures high reliability using both ARQ and FEC. Multiple access is provided by the distributed queuing request update multiple access (DQRUMA) protocol where PBSs control the bandwidth allocated to nodes. Nodes transmit reservation requests using a contention scheme. A unique feature of this protocol is that it allows a node to piggyback the transmission of ATM cells that are queued in the buffer. This helps to cut down the number of reservations needed to transmit ATM cells.

Currently, a project based on BAHAMA named Mobile Information Infrastructure (MII) is being carried out jointly by Bell Laboratories and Sun Microsystems and is partially supported by the U.S. National Institute of Standards and Technology (NIST). More information can be found in [10, 31].

6.5.3 Magic WAND

Magic WAND (Wireless ATM Network Demonstrator) is developed for customer premise networks under the Advanced Communications Technologies and Services (ACTS) program that is funded by the European Union. The project covers a whole range of functionality from basic wireless data transmission to shared multimedia applications.

Communication among mobile terminals and base stations takes place in the 5 GHz frequency band at a transmission speed of 20 Mbps and at a maximum range of 50 m. Higher data rate operation at 50 Mbps in the 17 GHz frequency band is also being studied.

The multiple access protocol (known as mobile access scheme based on contention and reservation for ATM or MASCARA) is a centrally controlled, adaptive TDMA/TDD scheme which combines reservation and contention to achieve efficient transmission and QoS guarantees. The traffic scheduling algorithm is delay-oriented and is designed to meet the requirements of different ATM service classes.

Field trials in hospital and office environments have been conducted to evaluate and demonstrate the technical feasibility of providing real-time multimedia services to mobile nodes.

More information can be found at http://www.tik.ee.ethz.ch/~wand.

6.5.4 MEDIAN

Another project under the Advanced Communications Technologies and Services (ACTS) program that is funded by the European Union. This project aims to demonstrate a 155 Mbps wireless ATM system operating in the 60 GHz band. The emphasis is on radio design and multiple access.

More information can be found at http://www.infowin.org/ACTS/PROJECTS.

6.5.5 AWACS

The Advanced Wireless ATM Communications Systems (AWACS) is a cooperative project between Europe and Japan. It is based on an 18 to 19 GHz wireless ATM LAN testbed with low-mobility multimedia terminals operating at data rates of up to 34 Mbps and at a range of up to 100 m. Possible use of the 40 GHz frequency band is also being studied. A virtual office field trial was successfully demonstrated. The base and mobile stations employ directional antennas and their locations were optimized through ATM cell error measurements.

Details of this project can be found at http://www.cselt.it/sonah/AWACS.

6.6 Commercial wireless ATM systems for local loops

Although wireless ATM is a relatively new concept, several ATM commercial systems have been developed for the wireless local loop. Among these systems include the Alcatel 9900 WW and the WI-LAN 300-24.

6.6.1 Alcatel 9900 WW

The Alcatel 9900 WW is a multiservice, point-to-multipoint wideband radio system that operates in frequency bands ranging from 10 to 41 GHz. The system is designed according to specifications set by the Digital Audio-Visual Council (DAVIC) Forum and ETSI. The product utilizes high-speed TDMA technology (with dynamic radio resource allocation for both packet and circuit modes) to relay wireless ATM cells within a wireless local loop. A central base station communicates with up to 3,000 nodes. The base station uses 7, 14, and 28 MHz TDM channels for downlink transmission while mobile nodes employ either 3.5 or 7 MHz TDMA channels for uplink transmission. A single omnidirectional antenna can be used. Alternatively, the wireless coverage area can be split into four regions, each served by a 90-degree antenna sector that operates on a different frequency channel. Each sector can support wireless data rates of 8, 16, and 34 Mbps using QPSK modulation with coding and interleaving. For four sectors, this means a maximum wireless data rate of 136 Mbps is available. The aim is to provide a fast roll-out of wireless infrastructure supporting broadband services (such as fast Internet, high-speed data, digitized voice, broadcast audio/video, video-on-demand) in small offices/home offices (SOHO), business enterprises, and residential areas. Operators can provide virtual dedicated circuits at rates ranging from $n \times 64$ Kbps to $n \times 2$ Mbps (which are equivalent to leased lines).

More information is available at http://www.alcatel.com.

6.6.2 WI-LAN 300-24

The WI-LAN 300-24 access point is another product designed for wireless local loop applications. It employs combined multicode DSSS and OFDM in the 2.4 GHz and 5.7 GHz ISM band. The polling-based multiple access technique dynamically allocates variable-length time slots with a peak data rate of 26 Mbps using 20 MHz of bandwidth. Like most wireless LANs, the 300-24 access point provides standard Ethernet interfaces with bridging features. Products with speeds in excess of 100 Mbps are currently being developed.

More information is available at http://www.wi-lan.com.

6.7 Satellite communications using ATM

Satellite ATM networks possess several advantages over terrestrial ATM networks. Satellites are relatively easy to deploy, are largely unobstructed, and are capable of providing large capacity broadcasts and global access to information services, even in sparsely-populated areas not economically serviceable by terrestrial radio coverage areas or cable networks. Moreover, satellite systems are virtually immune to impairments such as multipath because a signal propagating skyward does not encounter much reflection from surrounding objects.

In a 1997 study conducted by the FCC, it was reported that to link every home in the world to the Internet via optical fiber would cost US$300 billion [32]. To do the same with satellite coverage that spans the globe costs 30 times less or US$9 billion. Furthermore not all countries are connected via undersea optical fiber cables but most are linked to satellites, particularly satellites from the International Telecommunications Satellites (Intelsat) Organization.

Presently, satellites carry about a third of voice traffic and essentially all international television traffic [33]. Satellites are also starting to provide access to the Internet, intranets, and residential broadband services. By the beginning of the next millennium, the provision of broadband Internet services to vast regions are expected to become available [34]. In addition, satellites occupy an important role in broadband communications that involve a multitiered network approach. A system of low, medium, and geosynchronous orbit satellites may be envisioned as an overlay to the broadband terrestrial cellular network. However, new issues pertaining to satellite handoff, intersatellite links, multiple access, and error and congestion control will have to be identified and addressed.

6.7.1 Satellite networks

The design of satellite systems presents several architectural options, namely, low earth orbit (LEO), medium earth orbit (MEO), and geostationary earth orbit (GEO). These orbits affect the size of the geographical area (footprint) serviced by a satellite. Conventional GEO systems enjoy wide footprints but suffer from long propagation delays and high path loss. Thus, GEO satellites have been used to broadcast television signals and data whereas optical fiber cables are the preferred medium for carrying telephone voice signals. This scenario may change with the introduction of a completely new type of communication satellite that combines the benefits of cellular and satellite systems into a single global network. These LEO satellites are not only capable of amplifying and retransmitting signals but are also capable of switching and routing them [35]. LEO satellite systems employ a large number of satellites to achieve global coverage (see Figure 6.12). These satellites travel at a faster speed relative to the earth's rotation. Because the satellites are located much closer to the earth than GEO systems, LEO satellites introduce substantially less delays, incur low path loss, and require small antennas (see Table 6.6), attributes that suit the design of low-cost satellites as well as lightweight and portable handheld units. However, LEO systems do have their disadvantages. As each LEO satellite is visible to the ground terminal

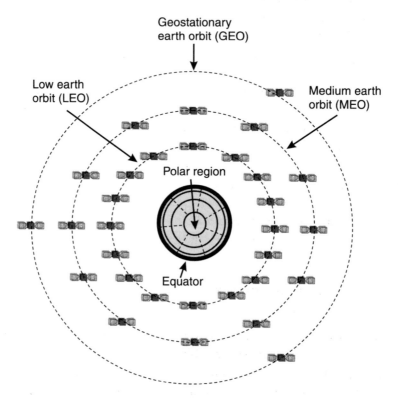

Figure 6.12   Satellite orbits.

Table 6.6
Characteristics of Satellite Orbits

Orbit	LEO	MEO	GEO
Altitude (km)	700 to 2000	7800 to 10300	35800
Number of Satellites	46 to 288	10 to 20	3 to 9
Roundtrip Delay (ms)*	10	80	250

*On-board processing and intersatelite links introduce additional delays.

for only a few minutes at a time, handoffs between satellites are frequent. Furthermore, acquiring and tracking these fast-moving satellites are nontrivial matters. Due to the high-speed movement of an LEO satellite relative to an observer on the earth, satellite systems using this type of orbit must also cope with large Doppler shifts. A Doppler shift causes the received signal to be frequency shifted with respect to the transmitted signal, a phenomenon that is proportional to the relative velocity and the frequency employed. MEO satellites offer features that represent a compromise between LEO and GEO systems.

Figure 6.13 illustrates the operation of an LEO satellite system. A spot beam services a geographical coverage area on earth, each about 150 km in diameter. Use of narrow spot beams is necessary because the signals from handheld portable devices are weak. Like cellular networks, clusters of these geographical areas enlarge the overall service region. If the

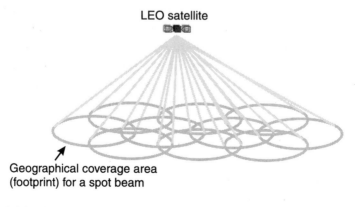

LEO satellite

Geographical coverage area
(footprint) for a spot beam

Figure 6.13   Spot beams of an LEO satellite.

duration of a connection is greater than the time the LEO satellite is sighted, the mobile user actually acts as a base station while the satellite serves as the mobile user.

6.7.2 Multiple access

A popular method for multiple access in satellite ATM systems is multi-frequency TDMA (MF-TDMA) as depicted in Figure 6.14. Each slot contains an ATM cell with extended header information containing forward error correction coding and in-band signaling. MF-TDMA allows on-demand bandwidth allocation. It reduces satellite antenna sizes and makes efficient use of transmission power. Each earth station may transmit on any frequency channel at a given time. Note that because of the large roundtrip time delay inherent in satellite systems and the relatively small delay spread, CDMA systems perform significantly worse than they do on terrestrial links.

6.7.3 Error control

Satellite networks typically employ concatenated codes (e.g., Reed Solomon codes with convolutional coding and Viterbi decoding for maximum-likelihood sequence estimation) to achieve high throughput while reducing antenna size and power requirements, thereby minimizing ground segment costs. Due to the channel coding, transmission errors in a satellite link are likely to occur in clusters (bursts). Consequently, multiple errors sometimes spread over both the payload and header of an ATM cell. These burst errors cannot be corrected by the HEC (since the HEC is capable of correcting only single bit errors) and may produce a high level of ATM cell loss and cell error. Even worse, certain burst error patterns can lead to undetected header errors that will result in the delivery of erroneous cells. Thus, burst errors are a major concern for adapting ATM technology over satellite. Selective interleaving is shown to be effective against error bursts (at the cost of increased delay) and can help to reduce significantly the ATM cell loss probability, as well as the probability of undetected error. An interleaver basically consists of a matrix with the number of rows determined by the expected maximum fade duration and the number of columns is equal to the decoding depth. The encoded data are written row by row and transmitted column by column. When interleaving and de-interleaving are applied to the ATM

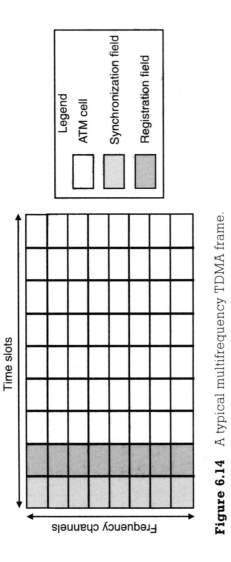

Figure 6.14 A typical multifrequency TDMA frame.

cell header and/or payload, error bursts become distributed or spread out in time. Since the burst errors now appear to be independent, they can be more easily corrected. Another novel method involves converting the HEC into one that can correct multiple bit errors. The header of a standard ATM cell comprises the 8-bit HEC plus 32 bits of header information. The approach described in Wu [32] creates a new 8-bit HEC based on the other 32 header bits. Thus, the overall ATM cell format remains unchanged. In addition, independence between the header and data fields as well as between ATM cells is preserved.

The impact of errors on the ATM layer over a satellite link has been investigated by several organizations. COMSAT tested a satellite ATM link from fractional T1 to T3 (45 Mbps) speeds [36] while EUTELSAT reported its results on 34 Mbps ATM links [37]. The COMSAT satellite system is based on Very Small Aperture Terminal (VSAT) technology. It employs a patented framing and interleaving algorithm. While performing interleaving, Reed Solomon forward error correction coding is added. The cell-based Reed Solomon codes adaptively apply 0 to 8% overhead depending on the measured satellite link quality. Coupled with lossless data compression ensures that satellite bandwidth is efficiently utilized. A number of pilot projects in various parts of the world are also in progress and new ones are planned to validate these results in operational environments. A key finding in these ATM testbeds is that a typical satellite link BER of roughly 10^{-7} yields an acceptable CLR of between 10^{-6} and 10^{-7}. Other major broadband satellite ATM projects include Aries in the United States and Canarie in Canada. These projects integrate satellite ATM with their national wireline ATM networks. NASA conducted several proof-of-concept trials and found satellite systems capable of delivering ATM cells at a rate of 622 Mbps. Emerging European projects on satellite ATM can be found in ETSI's TR 101 374-1[38]. Other broad-band satellite projects can also be found at http://www.atmdigest.com/ as well as references [39, 40].

6.7.4 Traffic and congestion control

Congestion is defined as a state in which the network is not able to meet the required QoS for currently established as well as new connections. A set of traffic control and congestion control procedures has been established for ATM. Traffic control refers to the actions taken by the network to avoid congested conditions. Congestion control refers to the actions taken by the network to minimize the intensity, spread, and duration of congestion. The relationship between these controls and the wireless environment has to be taken into consideration. Terrestrial ATM networks usually employ feedback mechanisms to minimize the probability of losses due to congestion. In a satellite environment, the large propagation delay prohibits the use of feedback methods. Some solutions suggested include controlling the cell loss priority (CLP) and employing backward explicit congestion notification (BECN) [41]. Other solutions require the use of buffering and virtual circuit prioritization. Generally, high priority traffic such as real-time traffic involving CBR or VBR video is subject to minimal buffering delay, thus guaranteed to maintain very good jitter performance. Low priority traffic such as nonreal-time ABR traffic can be transferred to large buffers.

6.7.5 Transport control

Without doubt, the Internet has dramatically changed the nature of global communications. The demand for Internet access from nearly all parts of the world is growing at an explosive rate. Satellite networks offer special advantages such as global reach, simplified network architecture, and asymmetric data rates. However, the maximum throughput is affected by the roundtrip propagation delay and the TCP window size. The maximum throughput can be estimated by the ratio of the maximum window size over the roundtrip time. For a maximum TCP window size of 65,535 octets (64K octets) and a geostationary satellite roundtrip time of 0.5 s, the maximum throughput is in the range of 1 Mbps. There is a proposal to increase the maximum TCP window size to 10^9 octets, potentially increasing the throughput by many times. Note that non-TCP applications are not limited by the 1 Mbps data rate. The user datagram protocol (UDP) is also not subjected to the throughput limit.

Slow start is a TCP/IP mechanism that can also affect the throughput of a satellite link significantly. Slow start limits the increase in the data rate to the same rate that acknowledgments are received, thereby protecting the Internet from network overload. The overall effect of the slow start algorithm is to increase the time required to reach the maximum throughput, both at the beginning of a TCP session and whenever a data segment is lost or corrupted. One of the remedies to recover from lost data segments quickly is to incorporate fast retransmit and fast recovery into the TCP protocol. The random early discard is a method that forces faster retransmission, rather than timeouts. It selectively penalizes high throughput TCP/IP sessions. Random early discard is expected to become widely deployed.

The service specific connection oriented protocol (SSCOP) is a transport protocol defined by the International Telecommunication Union-Telecommunication Sector (ITU-T) recommendation Q.2110 for reliable end-to-end delivery of data in ATM networks. Because SSCOP is designed to compensate for long propagation delays, it is well-suited for satellite ATM networks. SSCOP employs 24-bit sequence numbering and selective retransmission. The large sequence number space implies that a large window size is allowed and this can help to improve throughput performance. The selective retransmission mechanism prevents unnecessary retransmission, thus providing faster recovery of lost data.

The Internet Engineering Task Force (IETF) has chartered a working group to study the issues affecting TCP throughput over satellite links. More information can be found at http://www.ietf.org/html.charters/tcpsat-charter.html.

6.7.6   Intersatellite links

Most LEO satellite systems make use of direct intersatellite links (ISLs) to provide global connectivity, bypassing terrestrial (ground) networks and avoiding dependence on these resources. This facilitates provision of services to countries lacking in communications infrastructure. However, the use of ISLs increases delay variation and requires satellite on-board switching which increases processing delay. In addition, routing among these links is mandatory in order to exchange information with two or more distant nodes on earth. This entails identifying the start and end satellites and connecting these satellites via a network of ISLs. The mobility of the nodes and the movement of the satellites create a time-varying network topology. An approach in dealing with this problem is to first prepare a virtual topology based on virtual path (VP) connections and then search for available end-to-end routes within the ISL network [42].

6.7.7 Rain attenuation

Due to congestion on the frequencies currently used for fixed satellite services, the higher frequency Ka band (18 to 31 GHz) is emerging as the frequency of choice. The higher frequencies allow for narrow spot beams to be used, creating greater system capacity and bandwidth reuse. Since antenna size is largely a function of wavelength, which decreases as frequency increases, the antennas for satellite systems operating in the Ka band will be relatively small. While the Ka band provides ample bandwidth for delivery of multimedia services (recently, the ITU allocated 400 MHz of bandwidth for LEO or MEO fixed satellite systems), the main problem with this band is that signals are severely attenuated by rain. At these frequencies, the size of a raindrop becomes comparable to the wavelength. This causes Ka band signals to be weakened due to absorption and scattering when they travel through the raindrops. The problem is particularly widespread in tropical countries where the rate of annual rainfall is high. LEO satellite systems can mitigate rain fading effectively. The larger number of satellites can provide alternate paths (diversity) if signal conditions are poor due to rain.

6.8 Summary

ATM is rapidly becoming the base technology for next generation global communications, supporting diverse applications with different QoS requirements at a variety of speeds and distances. Using a connection-oriented approach and cell-based switching/multiplexing techniques, ATM allows voice, video, and data services to be carried transparently through a single integrated network. Wireless ATM exploits these capabilities with the added benefits of tetherless and location-independent access. It is an effective means to extend multimedia applications from the wireline to the wireless domain. However, due to the differences in characteristics between wireline and wireless links, the design of wireless ATM networks is aimed primarily at resolving issues related to mobility management and maintaining QoS in the presence of intermittent connectivity and variable bandwidth. Several research projects around the world have demonstrated the feasibility of deploying wireless ATM while several ATM products for the wireless local loop have been developed.

Satellite ATM will become increasingly important in providing worldwide multimedia services. The global reach of satellites is especially attractive for under-served remote regions or for parts of the world lacking in basic telecommunications infrastructure.

References

[1] Bing, B., "A Survey on Wireless ATM Technologies and Standardization," *Baltzer Journal on Telecommunication Systems,* Vol. 11, Nos. 3–4, April 1999, pp. 205–222.

[2] Kruys, J. "Wireless ATM—Tales of a Marriage," *Telecommunications,* February, 1997, pp. 39–46.

[3] *Realizing the Information Future: The Internet and Beyond,* Washington, D. C.: National Academy Press, 1994.

[4] *The Evolution of Untethered Communications,* Washington D.C.: National Academy Press, 1997.

[5] Shafi, M., et al., "Wireless Communications in the Twenty First Century: A Perspective," *Proceedings of the IEEE,* Vol. 85, No. 10, October 1997, pp. 1622–1638.

[6] The ATM Forum, "Real ATM is Here Today," *The ATM Forum Newsletter,* Vol. 4, No. 2, August 1996.

[7] National Science Foundation, Division of Networking and Communications Research and Infrastructure, "Research Priorities in Wireless and Mobile Communications and Networking," March 1997.

[8] Bing, B., and R. Subramanian, "Enhanced Reserved Polling Multiaccess Technique for Multimedia Personal Communication Systems," *ACM/Baltzer Journal on Wireless Networks,* Vol. 5, No. 3, May 1999, pp. 221–230.

[9] Tuch, B., "Development of WaveLAN, an ISM Band Wireless LAN," *AT&T Technical Journal,* Vol. 72, No. 4, July–August 1993, pp. 27–37.

[10] Ayanoglu, E., K. Eng and M. Karol, "Wireless ATM: Limits, Challenges and Proposals," *IEEE Personal Communications Magazine,* Vol. 3, No. 4, August 1996, pp. 18–34.

[11] Krishnakumar, A., "ATM Without Strings: An Overview of Wireless ATM," *Proceedings of the IEEE International Conference on Personal Wireless Communications,* 1996, pp. 216–221.

[12] Barry, J. et al., "High-speed Nondirective Optical Communication for Wireless Networks," *IEEE Network,* November 1991, Vol. 5, No. 6, pp. 44–54.

[13] Sanchez, J., R. Martinez and M. Marcellin, "A Survey of MAC Protocols Proposed for Wireless ATM," *IEEE Network,* Vol. 11, No. 6, November–December 1997, pp. 52–62.

[14] Acampora, A., "Wireless ATM: A Perspective on Issues and Prospects," *IEEE Personal Communications Magazine*, Vol. 3, No. 4, August 1996, pp. 8–17.

[15] Tobagi, F., R. Binder, and B. Leiner, "Packet Radio and Satellite Networks," *IEEE Communications Magazine*, Vol. 22, No. 11, November 1984, pp. 24–40.

[16] Morinaga, N., M. NaKagawa, and R. Kohno, "New Concepts and Technologies for Achieving Highly Reliable and High Capacity Multimedia Wireless Communications Systems," *IEEE Communications Magazine*, Vol. 35, No. 1, January 1997, pp. 34–40.

[17] Hui, J., *Switching and Traffic Theory for Integrated Broadband Networks*, Norwell, Massachusetts: Kluwer Publishers, 1990.

[18] Jacobs, I., R. Binder and E. Hoversten, "General Purpose Packet Satellite Networks," *Proceedings of the IEEE*, Vol. 66, No. 11, November 1978, pp. 1448–1467.

[19] Bing, B., "On the Stability of the Randomized Slotted ALOHA Protocol," *Proceedings of the IEEE International Symposium on Information Theory*, Cambridge, MA: Massachusetts Institute of Technology, August 1998, p. 223.

[20] Dellaverson, L., and W. Dellaverson, "Distributed Channel Access on Wireless ATM Links," *IEEE Communications Magazine*, Vol. 35, No. 11, November 1997, pp. 110–113.

[21] Cain, B., and D. McGregor, "A Recommended Error Control Architecture for ATM Networks with Wireless Links," *IEEE Journal on Selected Areas in Communications*, Vol. 15, No. 1, January 1997, pp. 16–28.

[22] Caceres, R., and L. Iftode, "Improving the Performance of Reliable Transport Protocols in Mobile Computing Environments," *IEEE Journal on Selected Areas in Communications*, Vol. 13, No. 5, June 1995, pp. 850–857.

[23] Balakrishnan, H., et al., "A Comparison of Mechanisms for Improving TCP Performance over Wireless Links," *Proceedings of ACM SIGCOMM*, August 1996, pp. 256–269.

[24] Acampora, A., "Strategic Plan 1996–2000," Center for Wireless Communications, University of California at San Diego, 1996.

[25] Perkins, C., *Mobile IP*, Reading, MA: Addison Wesley, 1998.

[26] Acharya, A., et al., "Mobility Management in Wireless ATM Networks," *IEEE Communications Magazine*, Vol. 35, No. 11, November 1997, pp. 100–109.

[27] Pancha, P., and M. Zarki, "Bandwidth Requirements of Variable Bit Rate MPEG Sources in ATM Networks," *Proceedings of the IEEE INFOCOM*, March 1993, pp. 902–909.

[28] Kwok, T., *ATM: A New Paradigm for Internet, Intranet and Broadband Services and Applications*, Englewood Cliffs, NJ: Prentice-Hall, 1998.

[29] Toy, M., *ATM Development and Applications*, New York: IEEE Press, 1996.

[30] Raychaudhuri, D., "WATMnet: A Prototype Wireless ATM System for Multimedia Personal Communication," *IEEE Journal on Selected Areas in Communications*, Vol. 15, No. 1, January 1997, pp. 83–95.

[31] Veeraraghavan, M., M. Karol and K. Eng, "Mobility and Connection Management in a Wireless ATM LAN," *IEEE Journal on Selected Areas in Communications*, Vol. 15, No. 1, January 1997, pp. 50–68.

[32] Wu, W., "Satellite Communications," *Proceedings of the IEEE*, Vol. 85, No. 6, June 1997, pp. 998–1010.

[33] Evans, J., "New Satellites for Personal Communications," *Scientific American*, Vol. 278, No. 4, April 1998, pp. 60–67.

[34] Mauger R., and C. Rosenberg, "QoS Guarantees for Multimedia Services on a TDMA-Based Satellite Network," *IEEE Communications Magazine*, Vol. 35, No. 7, July 1997, pp. 56–65.

[35] Wu, W., et al., "Mobile Satellite Communications," *Proceedings of the IEEE*, Vol. 82, No. 9, September 1994, pp. 1431–1448.

[36] Chitre, D., et al., "Asynchronous Transfer Mode (ATM) Operation via Satellite: Issues, Challenges and Resolutions," *International Journal of Satellite Communications*, Vol. 12, February 1994, pp. 211–221.

[37] Agnelli, S., and P. Mosca, "Transmission of Framed ATM Cell Streams over Satellite: A Field Experiment," *Proceedings of the IEEE International Conference on Communications*, June 1995, pp. 1567–1571.

[38] TR 101 374-1, "Satellite Earth Stations and Systems (SES); Broadband Satellite Multimedia; Part 1: Survey on Standardization Objectives," October 1998.

[39] Gilderson, J. and J. Cherkaoui, "Onboard Switching for ATM via Satellite," *IEEE Communications Magazine*, Vol. 35, No. 7, July 1997, pp. 66–70.

[40] Hung, A., M. Montepit and G. Kesidis, "ATM via Satellite: A Framework and Implementation," *ACM/Baltzer Journal on Wireless Networks*, Vol. 4, 1998, pp. 141–153.

[41] Akyildiz, I., and S. Jeong, "Satellite ATM Networks: A Survey," *IEEE Communications Magazine*, Vol. 35, No. 7, July 1997, pp. 30–43.

[42] M. Werner, et al., "ATM-Based Routing in LEO/MEO Satellite Networks with Inter-Satellite Links," *IEEE Journal on Selected Areas in Communications*, Vol. 15, No. 1, January 1997, pp. 69–82.

Contents

Wireless ATM Standardization

Wireless ATM standardization efforts are being carried out by the ATM Forum and the European Telecommunications Standards Institute (ETSI) Broadband Radio Access Network (BRAN) committee (formerly RES 10). To ensure compatible standards, there is close cooperation between the ATM Forum and BRAN. Both standards are expected to be completed by the year 2000. In Japan, the Multimedia Mobile Access Communication Systems Promotion Council (MMAC-PC) is also investigating high-speed wireless ATM access.

7.1 The ATM Forum wireless ATM standard

Fifty companies formed the wireless ATM working group of the ATM Forum in June 1996. The group is responsible for developing

a set of specifications that facilitates the deployment of ATM in both public
and private wireless access environments including wireless local loop,
personal communication systems, and satellite networks. The impetus to
establish a separate working group came about when some companies in
the radio industry realized that current radio transmission methods are
inefficient compared to ATM transmission over the air. In addition,
wireless ATM can help to extend ATM to the end-user.

The ATM Forum designed the wireless ATM specifications for compatibility with standard ATM protocols by providing ATM-based radio
access as well as extensions for mobility support within an ATM network
(see Figure 7.1). The system reference model (see Figure 7.2) consists of
a radio access segment and a fixed network segment. The fixed network
segment is defined by the "M" (mobile) user network interface (UNI) and
network node interface (NNI). The radio access segment is defined by the
"W" (wireless) UNI that includes the radio access layer (RAL).

The set of RAL protocols deals with physical radio transmission,
medium access control, data link (error) control, connection admission,
and radio resource management. Another set of mobility ATM (MATM)
protocol extensions is concerned with handoff signaling, location management for mobile terminals, routing and QoS provisioning for mobile
connections, and wireless network management. These protocols are
listed in Table 7.1.

Two location management schemes are proposed: location register
(LR) and private NNI (PNNI). In the LR scheme, a database is required to
register the location of the mobile terminal. This method is well-suited

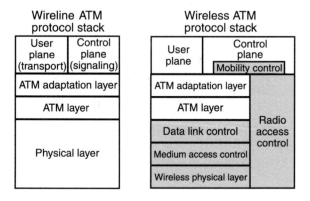

Figure 7.1   Wireline and wireless ATM protocol stacks.

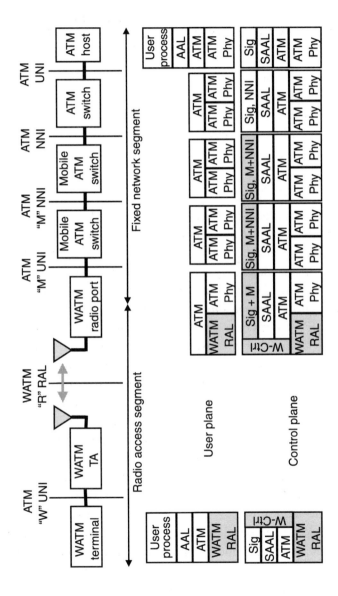

Figure 7.2 The ATM Forum's wireless ATM system reference model.

Table 7.1
Work Items of the ATM Forum's Wireless ATM Working Group

Radio Access Layer (RAL) Protocols	Mobile ATM (MATM) Protocol Extensions
Radio physical layer	Handoff signaling
Wireless medium access control	Location management
Wireless data link control	Connection routing
Radio resource assignment	Traffic and QoS control
Cell loss/sequencing during handoffs	Mobile network management

for highly mobile nodes. The PNNI scheme is based on the current PNNI for private ATM networks by extending signaling and routing protocols for location management. PNNI simplifies the configuration of large networks because it allows ATM switches to automatically learn about their neighbors and to distribute routing information dynamically. It embeds topological information in a hierarchical address structure that reflects the current network topology. Such hierarchical addressing enables PNNI to scale well for large ATM networks. PNNI operates on dedicated links but may also be tunneled over virtual path connections. Routing in PNNI is based on the well-known link-state routing technique employed by the open shortest path first (OSPF) IP routing protocol. PNNI is suitable for low mobility personal communication systems (PCS).

The MATM protocol extensions are independent of the radio subsystems and the frequency band employed. The RAL protocols, however, rely on the radio access technology and frequency band. It has been agreed that the RAL specifications using the 5 GHz band (and providing a capacity in the region of 25 Mbps) be developed first since this band is available for high-speed wireless access in both America and Europe. For example, the high-speed Unlicensed National Information Infrastructure or U-NII (formerly NII/SUPERNET) is the next generation of wireless information transmission systems to be deployed in the United States. Wireless access through the U-NII is an inexpensive method of interconnecting schools, libraries, and hospitals. The large amount of radio spectrum allocated (300 MHz) enables the provision of high-speed Internet and multimedia services.

The total U-NII bandwidth allocation is broken down into three unchannelized blocks of 100 MHz, each with a power limit that depends on the range supported (see Table 7.2). Since the spectrum allocated is not licensed, large-scale frequency planning is avoided and ad-hoc networks are possible. However, due to the difficulty in controlling interference, nodes must observe etiquette during transmission so that incompatible systems can coexist. The key elements of etiquette are:

▶ Listening before transmitting;

▶ Limiting transmission time;

▶ Limiting transmit power.

In order to achieve transparent access between wireline and wireless ATM networks, the peripheral devices for wireless ATM will be based on standard ATM protocols. Wireless ATM can also occur over wireless networks based on IP or PCS. Protocol conversion is necessary to internetwork ATM with these networks. The first U-NII products will probably be in the 5.25 to 5.35 GHz band because the power levels and frequency bands are the same as HIPERLAN. More details on ATM Forum's activities can be found at http://www.atmforum.com.

7.2   The BRAN standards

ETSI's work on radio networks is classified under the generic name of High Performance Radio LANs (HIPERLAN). It was initiated by the former Radio Equipment and Systems (RES10) technical committee which developed the HIPERLAN Type 1 functional specification. ETSI has recently

Table 7.2
Characteristics of U-NII Frequency Bands

Frequency Band	Power Limit	Range
5.15 to 5.25 GHz	200 mW EIRP	Short
5.25 to 5.35 GHz	1 W EIRP	Medium
5.725 to 5.825 GHz	4 W EIRP	Long

established a new standardization project for Broadband Radio Access Networks (BRAN). This project provides facilities for high-speed access to wired and wireless networks in both business and residential premises. To ensure overall coherence with other existing and emerging technologies, close relationships have been or are being established with the ATM Forum, the IEEE 802.11 Wireless LAN committee, the Internet Engineering Task Force (IETF), the MMAC-PC, the International Telecommunication Union-Radio Sector (ITU-R), and a number of internal ETSI technical bodies.

The BRAN committee is mainly looking at the frequency band of 5.15 to 5.30 GHz (which is channelized at 23.5 MHz) for both indoor and remote wireless ATM access at roughly 25 Mbps. Another band of interest is the 17.1 to 17.3 GHz band, which is being targeted for wireless ATM interconnection with data rates in excess of 150 Mbps. The BRAN committee is currently developing standards for the following types of high-speed radio access networks:

- HIPERLAN/2—This short range (up to 50 m) 5 GHz standard offers up to 25 Mbps high-speed access to a variety of wireless networks including those based on UMTS, ATM, and IP.

- HIPERACCESS—This long range (up to 5 km) standard offers up to 25 Mbps high-speed access to a variety of wireless networks including those based on UMTS, ATM, and IP. Spectrum allocations are currently being decided and may range from 2 to 60 GHz.

- HIPERLINK—This short range (up to 150 m) 17 GHz standard offers up to 155 Mbps and is intended for the interconnection of HIPER-LANs and HIPERACCESS networks. About 200 MHz of bandwidth has been allocated in the 17 GHz band.

The specifications developed will address both the physical and data link layers (see Table 7.3). Interworking specifications that allow broadband radio systems to coexist with wired networks (such as those based on the Universal Mobile Telecommunications System or UMTS, ATM, and IP) will also be established. The relationship between HIPERLAN and UMTS is shown in Figure 7.3. UMTS (or IMT-2000 as it is known outside Europe) supports a mobile data rate of 144 Kbps, a portable data rate of

Table 7.3
An Overview of HIPERLAN, HIPERACCESS, and HIPERLINK Standards

Parameter	HIPERLAN Type 1	HIPERLAN Type 2	HIPERACCESS	HIPERLINK
Network Type	Wireless 8802 LAN	Wireless IP and ATM Short Range Access	Wireless IP and ATM Remote Access	Wireless Broadband Interconnect
Typical Operating Environment	Indoors	Indoors	Outdoors	Indoors/ Outdoors
Interface	Medium Access Control	Data Link Control	Data Link Control	Data Link Control
Frequency Band	5 GHz	5 GHz	Currently Not Available	17 GHz
Maximum Data Rate	23.5 Mbps	25 Mbps	25 Mbps	155 Mbps
Maximum Transmit Power	1 W	1 W	Currently Not Available	100 mW

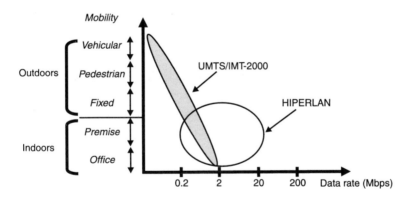

Figure 7.3   Relationship between HIPERLAN and UMTS.

384 Kbps, and a fixed rate of 2 Mbps for in-building communications. It is primarily geared towards voice, data, and low-quality video services. The work on ATM is done in cooperation with the ATM Forum which will focus on the signaling aspects associated with wireless ATM. A common reference model for BRAN supporting ATM (also known as

Wireless ATM Access Systems or WACS) has been defined. More details can be found in [1, 2], and http://www.etsi.org/bran/bran.htm.

7.3 The MMAC-PC

The MMAC-PC was formed in 1996 to propose a high performance wireless system to be used after IMT-2000. Four major specifications targeted by the council include:

 ▶ High-Speed Wireless Access—A 500 MHz to 1 GHz bandwidth mobile communication system which can transmit at up to 30 Mbps using the 25, 40, and 60 GHz bands. It is meant for mobile video telephone applications.

 ▶ Ultra High-Speed Wireless LAN—A 1 to 2 GHz bandwidth wireless LAN which can transmit at up to 156 Mbps using the 60 GHz millimeter wave band. High-quality TV conferencing is a typical application.

 ▶ 5 GHz Band Mobile Access—An ATM or Ethernet wireless LAN based on the 5 GHz band. Either system can transmit at up to 20 to 25 Mbps.

 ▶ Wireless Home-Link—An indoor wireless link which can transmit at up to 100 Mbps using the 5, 25, 40, and 60 GHz bands. It can be used for data transfer between PCs as well as between audio-visual equipment.

More information can be found in http://www.arib.or.jp/mmac/e/index.htm.

7.4 Summary

Wireless ATM standardization activities in the ATM Forum and the ETSI BRAN committee are making good progress. Although a universal wireless ATM standard in the 5 GHz frequency range is crucial for future mass market success, it is clear that commercial wireless ATM systems can become popular in the near future.

References

[1]   TR 101 031, "Broadband Radio Access Networks (BRAN); High Performance Radio Local Area Network (HIPERLAN) Type 2; Requirements and Architectures for Wireless Broadband Access," January 1999.

[2]   TR 101 378, "Broadband Radio Access Networks (BRAN); Common ETSI-ATM Forum Reference Model for Wireless ATM Access Systems (WACS)," December 1998.

Glossary

AAL ATM Adaptation Layer

ABR Available Bit Rate

ACK Acknowledgment

ACTS Advanced Communications Technologies and Services

ARQ Automatic Repeat Request

ATM Asynchronous Transfer Mode

BCH Bose-Chaudhuri-Hocquenghem

BER Bit Error Rate

BPSK Binary Phase Shift Keying

BRAN Broadband Radio Access Network

BSA Basic Service Area

BSS   Basic Service Set

BSSID   BSS Indentification

BT   Bandwidth-Time product

CAC   Channel Access Control

CBR   Constant Bit Rate

CCA   Clear Channel Assessment

CCK   Complementary Code Keying

CDMA   Code Division Multiple Access

CDV   Cell Delay Variation

CEPT   European Conference of Postal and Telecommunications Administrations

CGI   Common Gateway Interface

CLP   Cell Loss Priority

CLR   Cell Loss Ratio

CM   Control Module

CPFSK   Continuous Phase Frequency Shift Keying

CPM   Continuous Phase Modulation

CRC   Cyclic Redundancy Check

CSMA   Carrier Sense Multiple Access

CSMA/CA   CSMA with Collision Avoidance

CSMA/CD   CSMA with Collision Detection

CTD   Cell Transfer Delay

CTS   Clear to Send

DAVIC   Digital Audio-Visual Council

DBPSK   Differential Binary Phase Shift Keying

DCF   Distributed Co-ordination Function

DCLA   Direct Current Level Adjustment

DECT   Digital Enhanced Cordless Telecommunications

DFE   Decision Feedback Equalizer

DFIR   Diffuse Infrared

DHCP   Dynamic Host Configuration Protocol

DIFS   DCF Interframe Space

DQPSK   Differential Quaternary Phase Shift Keying

DS   Distribution System

DSL   Digital Subscriber Line

DSSS   Direct-Sequence Spread Spectrum

DTIM   Delivery TIM

EIRP   Equivalent Isotropically Radiated Power

ESS   Extended Service Set

ESSID   ESS Identification

ETSI   European Telecommunications and Standard Institute

FCC   Federal Communications Commission

FCS   Frame Check Sequence

FDD   Frequency Division Duplex

FDMA   Frequency Division Multiple Access

FEC   Forward Error Correction

FHSS   Frequency Hopping Spread Spectrum

GEO   Geostationary Earth Orbit

GFC   Generic Flow Control

GFSK   Gaussian Frequency Shift Keying

GMSK   Gaussian Minimum Shift Keying

HEC   Header Error Control

HIPERLAN   High Performance LAN

HTML   Hyper-Text Markup Language

IBSS   Independent Basic Service Set

IEC   International Electrotechnical Commission

IEEE   Institute of Electrical and Electronic Engineers

IETF   Internet Engineering Task Force

IFS   Interframe Space

IMT-2000   International Mobile Telecommunications 2000

INTELSAT   International Telecommunications Satellites Organization

IP   Internet Protocol

IrDA   Infrared Data Association

ISA   Industry Standard Architecture

ISDN   Integrated Service Digital Network

ISL   Inter-Satellite Link

ISM   Industrial, Scientific, and Medical

ISO   International Standards Organization

ITU-T   ITU-Telecommunications Sector

LAN   Local Area Network

LEO   Low Earth Orbit

LLC   Logical Link Control

MAC   Medium Access Control

MEO   Medium Earth Orbit

MF-TDMA   Multi-Frequency TDMA

MIB   Management Information Base

MKK   Musen-setsubi Kensa-kentei Kyokai

MMAC-PC   Multimedia Mobile Access Communication Systems Promotion Control

MPDU   MAC Protocol Data Unit

MPEG   Motion Pictures Expert Group

MSK   Minimum Shift Keying

NASA   National Aeronautics and Space Administration

NAV   Network Allocation Vector

NDIS   Network Driver Interface Specification

NIC   Network Interface Card

NNI   Network-Network Interface

NRL   Normalized Residual Lifetime

OFDM   Orthogonal Frequency Division Multiplexing

OSI   Open System Interconnection

OSPF   Open Shortest Path Forwarding

PC   Personal Computer

PCCA   Portable Computer and Communications Association

PCF   Point Coordination Function

PCI   Peripheral Component Interconnect

PCM   Pulse Code Modulation

PCS   Personal Communications System

PHY   Physical

PIFS   PCF Interframe Space

PLCP   Physical Layer Convergence Protocol

PMD   Physical Medium Dependent

PNNI   Private NNI

PPDU   PLCP Protocol Data Unit

PPM   Pulse Position Modulation

PT   Payload Type

QAM   Quadrature Amplitude Modulation

QoS   Quality of Service

QPSK   Quaternary Phase Shift Keying

RS   Reed Solomon

RTS   Request to Send

SFD   Start Frame Delimiter

SIFS   Short IFS

SOHO   Small Office Home Office

STP   Shielded Twisted Pair

SWAP   Shared Wireless Access Protocol

TCP   Transmission Control Protocol

TDD   Time Division Duplex

TDMA   Time Division Multiple Access

TIM   Traffic Indication Map

UBR   Unspecified Bit Rate

UM   User Module

UMTS   Universal Mobile Telecommunications Systems

UNI   User Network Interface

U-NII   Unlicensed National Information Infrastructure

USB   Universal Serial Bus

UTP   Unshielded Twisted Pair

VBR   Variable Bit Rate

VC   Virtual Circuit

VCI   Virtual Circuit Identifier

VP   Virtual Path

VPI   Virtual Path Identifier

VSAT   Very Small Aperture Terminal

WAP   Wireless Application Protocol

W3C   World Wide Web Consortium

WDF   Wireless Data Forum

WECA   Wireless Ethernet Compatibility Alliance

WEP   Wired Equivalent Privacy

WLANA   Wireless LAN Alliance

WLIF   Wireless LAN Interoperability Forum

WWW   World Wide Web

XML   Extended Markup Language

About the author

Benny Bing was formerly with the Department of Electronic and Computer Engineering at Ngee Ann Polytechnic in Singapore where he specialized in data communications and computer networking. Currently, he is with the Department of Electrical and Computer Engineering at the University of Maryland in College Park, Maryland, USA. He participated in various satellite communications and networked video conferencing projects while working with Singapore Telecom and AT&T Global Information Solutions. He was also active in the area of wireless communications research, contributing close to 20 refereed papers in reputable journals and conferences. His paper on wireless ATM was recently judged to be one of the best papers at the 1998 IEEE International Conference on ATM. He has chaired technical sessions, conducted tutorials, and delivered plenary lectures at IEEE international conferences. He served as a member of the international advisory committee for the 1999 International Conference on ATM and is a member of the joint technical program committee for the IEEE ATM Workshop 2000 and the 2000 International Conference on ATM. He is guest editor on Multiple Access for Broadband Wireless Networks for the *IEEE Communications Magazine.* He holds a B.Eng (Honors) degree and a M.Eng degree (by research) from the Nanyang Technological University in Singapore, all in Electrical and Electronic Engineering. He is a member of the IEEE and the IEEE Communications Society. Besides his wide interest in communication networks, he is also an avid fan of human computer interface and graphic design. He can be reached at bennybing@onebox.com.

Index

Wideband CDMA for Third Generation Mobile Communications,
Tero Ojanperä and Ramjee Prasad, editors

*Wireless Communications in Developing Countries: Cellular and
Satellite Systems,* Rachael E. Schwartz

Wireless Technician's Handbook, Andrew Miceli

For further information on these and other Artech House titles,
including previously considered out-of-print books now available
through our In-Print-Forever® (IPF®) program, contact:

Artech House
685 Canton Street
Norwood, MA 02062
Phone: 781-769-9750
Fax: 781-769-6334
e-mail: artech@artechhouse.com

Artech House
46 Gillingham Street
London SW1V 1AH UK
Phone: +44 (0)20 7596-8750
Fax: +44 (0)20 7630-0166
e-mail: artech-uk@artechhouse.com

Find us on the World Wide Web at:
www.artechhouse.com